ADDITION

THE AI

"A must-read for airplane buffs."

—Maj. Gen. William A. Anders, USAF (ret.), former Apollo 8 astronaut and president of the Heritage Flight Museum

"The intersection of engineering and ideas: that is where Jay Spenser believes you will find the real magic of flight. Spenser loves airplanes and the people that make them. His new book asks a very simple question—Why do we have the airplanes that we do?—that allows us to think about a familiar story in new ways. This celebration of progress offers us a fascinating way to think about the evolution of key ideas that define the form and function of the modern airplane."

—Dr. Deborah G. Douglas, Curator of Science and Technology, MIT Museum

"A fast-paced, informative overview."

—*Seattle Post-Intelligencer*

"*The Airplane* presents an insightful and innovative approach to some of the most intriguing questions in aviation. Jay Spenser has brilliantly demonstrated his mastery of aviation history by cleverly integrating disparate themes and ideas into a coherent and highly readable text that provides a clear explanation of complex aeronautical technology."

—Dr. F. Robert van der Linden, Chairman, Aeronautics Division, National Air and Space Museum, Smithsonian Institution

"Reading Jay Spenser's wonderful contribution to aviation history is akin to having a pleasant conversation over coffee, but without missing any of the detail and connections that any self-respecting historian should bring to the table. This book will bring aviation history alive for new generations of readers, especially our younger folks who have not yet learned the very personal joys of flying and of airplanes themselves."

—Dan Hagedorn, Senior Curator,
the Museum of Flight

"A smart . . . history of a thrilling machine all too often taken for granted."

—*Publishers Weekly*

"Jay Spenser tells the fascinating story of the development of heavier-than-air flight in an easy-to-read conversational style. Anyone interested in airplanes, from flying to technical design, will find this book a valuable addition to his or her library."

—Dr. John Anderson, Professor Emeritus,
Aerospace Engineering, University of Maryland

"An engaging text. . . . The lively writing and the number of photographs set it above many of its competitors."

—*Library Journal*

PHOTOGRAPH BY DEBORAH SPENSER

About the Author

JAY SPENSER has spent a lifetime studying aviation as a museum curator at the National Air and Space Museum and the Museum of Flight, and subsequently as an aerospace industry writer. He is the co-author of 747 and lives in Seattle, Washington.

THE AIRPLANE

Smithsonian Books

HARPER

NEW YORK • LONDON • TORONTO • SYDNEY

HOW IDEAS GAVE US WINGS

JAY SPENSER

HARPER

A hardcover edition of this book was published in 2008 by Collins.

FIRST HARPER PAPERBACK PUBLISHED 2009.

Designed by Jessica Shatan Heslin/Studio Shatan, Inc.

Cover photographs:
Top: Passengers board a Sikorsky S-40 in the early 1930s.

Bottom: Henri Farman completes Europe's first circling flight to land where he took off, January 1908.

Frontispiece: Hubert Latham sets off across the English Channel, July 1909.

The Library of Congress has catalogued the hardcover edition as follows:

Spenser, Jay P.
The airplane : how ideas gave us wings / Jay Spenser.
p. cm.
Includes bibliographical references.
ISBN 978-0-06-125919-7
1. Airplanes—Design and construction—History. 2. Aeronautics—Research—History.
I. Title.
TL671.2.S67 2008
629.133'34—dc22

2008023423

ISBN 978-0-06-125920-3 (pbk.)

HB 02.09.2023

To Donald S. Lopez, 1923–2008

CONTENTS

INTRODUCTION

THE QUEST FOR WINGS—HUMANKIND'S OLDEST DREAM

In New Hampshire one morning when I was seven years old, I awoke and found I could fly. Concentrating hard, I levitated off the floor, flew downstairs, rounded a tight corner, and soared across the living room of my grandparents' farmhouse without ever touching down.

It was a dream, of course, but it thrilled me, and decades later the lingering memory remains vivid. It turns out that I'm not alone; at one time or another in our lives, most of us have fantasized about flying.

Flying is humankind's oldest dream. Ever since our earliest ancestors first gazed skyward, we human beings have envied the birds their ability to slip gravity's bonds. The result is a pan-cultural longing for wings so deeply rooted in our psyche that it sings to our soul and is perennially our favorite metaphor for freedom.

The power of flight's siren call is difficult to overstate. Eons before a Hellenistic sculptor crafted the Nike of Samothrace—antiquity's great expression of this wish—flying was already a favored theme of artists and storytellers. From the dawn of history, it has colored our myths and our magical thinking. In religions, the ability to fly is universally equated with the divine.

As human technological prowess advanced, this yearning escalated into a focused quest for wings that ultimately succeeded because of people like you and me—dreamers all, and all mere mortals of flesh and blood. It is to their vision, ingenuity, collaboration, and sacrifice that we owe the modern wonder of air travel.

Inventing the airplane is one of history's greatest adventures, yet books about flight seldom do the subject justice. Too often they fail to evoke the underlying wonder of the subject matter. Worse still, air travel itself is often little fun these days. Long lines at airport security checkpoints, crowded flights, and today's lack of in-flight service and amenities have squeezed the glamour out of what should by all rights be an exciting adventure.

As for the jetliners themselves, we take them for granted. Whether we as individuals love to fly or dread it, we tend to view the airliners we board as strictly a conveyance—glorified buses with wings—and thus fail to see them for what they really are: an invention bequeathing a degree of mobility utterly inconceivable through the vast majority of human existence.

But what if we could see the jetliner with fresh eyes? Better still, what if we could understand the great ideas in history that laid its keel and sent it soaring high into the blue? Best of all, what if we could stand elbow to elbow with flight's pioneers and share vicariously in those "aha" moments when they solved aviation's technical challenges? Then flying would be an adventure again.

In this new century marked by profound global challenges, may this book remind us what we humans can do when we share a grand vision and collaborate to see it achieved.

JAY SPENSER

SEATTLE, WASHINGTON

MARCH 2008

1 CONCEPTION

THE THINKER AND THE DREAMER

An uninterrupted navigable ocean that comes to the threshold of every man's door ought not to be neglected as a source of human gratification and advantage.

—SIR GEORGE CAYLEY (1773–1857)[1]

In Yorkshire in the northeast of England, a human being first imagined the airplane. This scientifically accurate emergence happened a hundred years before the Wright brothers invented the real thing.

At first glance, Yorkshire seems an odd place for the science of aviation to begin. However, history shows that creativity flourishes where cultures mix, and England's largest traditional county certainly boasted plenty of that. Celtic tribes lost to the mists of time, marching Roman legions, Angle farmers settling from Germany, and marauding Vikings invading from Denmark all called it home at one time or another.

The airplane's conceptual inventor was Yorkshire baronet Sir George Cayley. Born in December 1773 at Scarborough on the North Sea, Cayley inherited his title, wealth, and large landholdings upon the death of his father. But a greater inheritance had already come his way at birth, for he possessed a brilliant mind.

Few people today know Cayley's name even though he single-handedly established the science of aviation and laid a foundation for others to build on. The Wright brothers never would have left the ground without his powerful ideas, for example, but they were far from the first to try.

Sir George Cayley.

That honor belongs to another Englishman, Cayley's self-appointed disciple William S. Henson. Thrilled by Cayley's visionary writings, Henson galloped off to design a real airplane before the middle of the nineteenth century. Although his premature attempt failed, Henson at least showed the world what the airplane would be.

If Cayley was the thinker, Henson—four decades his junior—was the dreamer. The two men hardly could have been more different, yet their overlapping efforts synergistically planted the seeds of flight.

The people of Yorkshire are known for a calm and deliberate nature. George Cayley from an early age broke the mold. Around his tenth birthday, this enthusiastic young aristocrat was excited in particular by news sweeping England: human beings had flown in Europe.

On November 21, 1783, Jean-François Pilâtre de Rozier and the marquis d'Arlandes ascended into the heavens in a new invention called the balloon. According to the reports, these Frenchmen drifted over the city of Paris for twenty-five minutes, covering 5½ miles (9 kilometers) before setting down safely.

At that time, the event was hailed as the first time human beings had ever flown. Today we know this was probably not the case. While history does not provide definitive proof of earlier manned ascents, it is quite likely that large kites (a dangerous way to fly, given their propensity for headlong plunges) carried people aloft more than a millennium before the invention of the balloon. The Venetian Marco Polo lends credence to accounts of earlier aerial forays. Writing in the late thirteenth century, he described personally witnessing people flying aboard large kites in China.

Pilâtre de Rozier and Arlandes' vehicle of 1783 was the brainchild of Joseph and Étienne Montgolfier, two brothers in France's papermaking trade. A majestic blue orb of varnished taffeta decorated ornately in gold, this hot-air balloon was open at the bottom and was launched after being filled with smoke from a large outdoor blaze before its restraints were released.

Surprisingly, the Montgolfiers did not know *why* their balloon sailed into the sky. They did not understand that hot air has a lower density than cold air and is thus lighter, so they instead endorsed the classical notion that it was smoke's natural tendency to rise that made their invention buoyant. Lending pseudoscientific credence to this flawed theory, they further asserted that smoke contained a previously unidentified substance—called *Montgolfier gas*, naturally—that imparts a gravity-defying upward force called *levity*.

Their success—and that of their archrival, French physicist Jacques Alexandre Charles with his more advanced hydrogen balloons—launched a rapturous, all-out French obsession with lighter-than-air flight. Part of this euphoria was the uplifting grace of balloons themselves, which lyrically fulfilled humankind's age-old dream of flight.

But there was more to this rampant "balloonacy" than poetic sensibilities. With the industrial revolution then under way in England and spreading to Europe, balloons also symbolized man's growing technological prowess and the heady excitement of new frontiers. Balloons even became a favorite decorative motif in French furniture, plates, paintings, mantel clocks, and chandeliers.

Back in Yorkshire, the success of the Montgolfiers kindled in young George Cayley a lifelong fascination with flight. But the balloon itself didn't hold the Yorkshire boy's interest for very long. He quickly decided that heavier-than-air vehicles were flying's future.

Two factors shaped this conviction. The first was Cayley's belief that a flying machine, to be practical, must be *dirigible* (steerable) so people could fly it where they liked instead of drifting at the whim of the wind. The second was his delight in a flying toy perfected a year after that first balloon flight by two other Frenchmen, the naturalist Launoy and a mechanic named Bienvenu.

Launoy and Bienvenu's toy was a rudimentary helicopter with a central shaft, corks at both ends with feathers angled to provide lift as they spun, and a bow (as in bow and arrow) drawn taut by winding its string around the shaft. Letting go the wound-up helicopter released the bow's tension, rotating the feathered shaft to carry it high into the sky.

In his early twenties, Cayley built and tested a copy of this ingenious device, which for him was more than a mere amusement. In size and performance, it greatly improved on the Chinese top, that ancient and ubiquitous toy consisting of a carved propeller mounted atop a stick. Spinning this stick rapidly between one's hands would send the Chinese top aloft.

Unlike balloons, man-made amusements such as these were not buoyant. Neither were birds, yet they too could fly. Such being the case, Cayley wondered, why couldn't a man-carrying machine be built that likewise was heavier than the air around it?

To investigate this intriguing idea, Cayley created a laboratory-cum-workshop at Brompton Hall, his ancestral estate at Brompton-by-Sawdon, near Scarborough. There he built models that he dropped down the manor house's stairwell in order to study their fall. His wife's tolerance of these highly disruptive experiments unfortunately proved low, so he conducted them only when she was away.

By 1799, George Cayley's pioneering efforts led him in his mid-twenties to an astonishing conceptual leap: the first scientifically grounded imagining of an airplane. That same year, the French Revolution drew to a close and Napoleon Bonaparte, the general who would be emperor, marched off to begin changing the face of Europe. George Washington died at his Virginia farm at age sixty-seven, the Rosetta Stone was discovered in Egypt, and Ludwig van Beethoven—not yet thirty and already going deaf—was at work on his first symphony.

On a silver disc dated 1799, Cayley inscribed a flying vehicle with an arched main wing, a single-seat gondola, and a tail resembling an arrow's stabilizing feathers. Attached by a universal joint, this cruciform tail could tilt up, down, or side to side to alter the craft's direction of flight.

The wing of this proto-airplane was a billowing fabric sail that Cayley apparently proposed with ease of construction in mind. Later in life in a second round of aeronautical experimentation, he would construct manned gliders with fabric wings.

The final feature of this crude etching reveals Cayley's greatest realization. Aft of the wing are propulsive paddles worked like oars by the pilot in the cockpit. Cayley called these paddles *propellers* even though they moved fore and aft rather than rotating.

Cayley understood full well that this fanciful propulsion system would not work and that human muscle power alone would be inadequate to sustain flight. He included this representation of a rowing impetus only as a placeholder for some future propulsion system that would include a *first mover*, as he called the concept of a mechanical engine.

Steam was then beginning to power England's industrial revolution, but steam engines were too heavy to fly. Consequently, Cayley contended his entire life with the frustrating lack of a suitable power plant. This was the single disappointment in an otherwise astonishingly successful career in the branch of science that he founded.

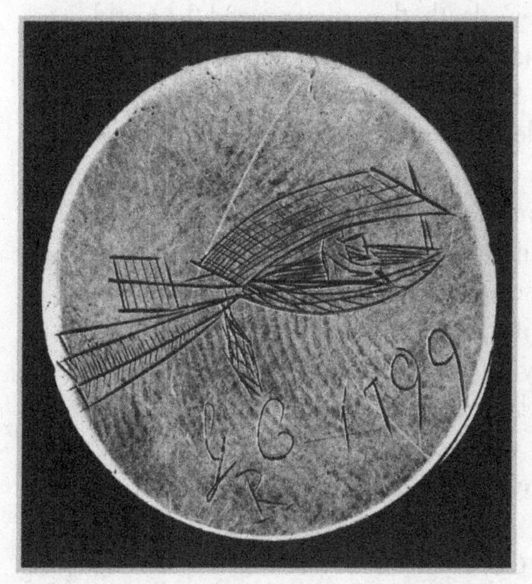

A century before the Wright brothers, George Cayley etched onto this silver disc history's first-ever imagining of the airplane.

For humans to fly, Cayley correctly observed, "it is only necessary to have a first mover that will generate more power in a given time, in proportion to its weight, than the animal system of muscles."[2] The development and successive refinement of internal-combustion gasoline engines during the nineteenth century would provide the Wright brothers and other pioneers with this last missing bit of technology by the start of the twentieth century.

Why is it so significant that Cayley called for a separate propulsion system? Because before he came along, people drew the wrong lesson from nature. They assumed that airplanes, if and when such things were invented, must achieve flight by flapping their wings like birds. We call this flapping-wing aircraft an *ornithopter.*

The trouble with ornithopters is that they are mechanical nightmares. Models can fly this way, but not full-size aircraft because of the enormous complexity and high stresses of ornithopter flight. To get a sense of just how unwieldy a concept this is, try to imagine a flapping-wing jetliner.

We have the advantage of hindsight, but at the time ornithopters looked like not just the logical way to fly but the *only* way. Leonardo da Vinci thought so, and he arguably possessed the greatest mind of the Renaissance. When his notebooks were rediscovered and published decades after Cayley's death, they mesmerized the world.

Sprinkled among Leonardo's thirteen hundred pages of handwritten notes are more than a hundred drawings that show his fascination with birds and remarkable concepts for human flight. Among these are beautiful illustrations of flapping-wing machines (ornithoptering hang gliders) suggestive of the anatomy of bats and birds. Da Vinci also sketched an unworkable vertical-flight machine that drew inspiration from Archimedes' screw and anticipated the helicopter.

Like da Vinci three centuries earlier, George Cayley took his inspiration from nature. Studying birds closely, he gleaned insights from their shapes, their flight, and how their wings were articulated and moved. In particular, he was fascinated to note that seagulls, when riding the winds of England's rugged north coast, soared for extended periods without flapping.

Mulling this over gave Cayley an "aha" moment that was fledgling aviation's single most crucial epiphany: flight is possible with wings that are held entirely rigid.

This was excellent news indeed. It meant that heavier-than-air flying machines could be designed that would be vastly simpler to construct, operate, and maintain than ornithopters. Yes, rigid wings would do the trick—so long as air flowed over them.

During a descent with gravity tugging, air would of course flow over a flying machine's wings. But how do you climb or sustain level flight when gravity must be defied? In these challenging flight phases, an impetus would be required to thrust the machine through the air.

This forward thrust had to come from somewhere, Cayley reasoned, and if the wings aren't flapping, then something else aboard the airplane must provide the push. That in turn dictated a separate propulsion system.

In one fell swoop, young Cayley sidestepped the ornithopter trap that even da Vinci had fallen into. Cayley's greatest legacy would be the layout for a powered, heavier-than-air flying machine whose wings do not flap because they are called on to provide only lift, not lift and thrust combined. Conceptually, the airplane now existed.

Prescient as he was, Cayley missed the bull's-eye in two regards with his 1799 design. One was his avoidance of rotating propellers even though the concept of the airscrew, or propeller, had been known to him since 1796. Another was his provision for control around two axes but not the third (the Wright brothers would address both these forgivable failures). In at least one key regard, however, Cayley's thinking outstripped that of the Wrights a century later: Cayley placed the airplane's elevator, the movable surface that tilts an airplane up or down, at the rear, not the front.

On the other side of his silver medallion, the twenty-six-year-old inventor identified gravity, lift, drag, and thrust as the forces involved in flight. In so doing, he showed himself to be the first person ever to properly understand flight's underlying scientific principles. "The whole problem," he later wrote, "is confined within these limits, *viz.* to make a surface support a given weight by the application of power to the resistance of air."[3]

For Cayley, this 1799 breakthrough was just the beginning. Where

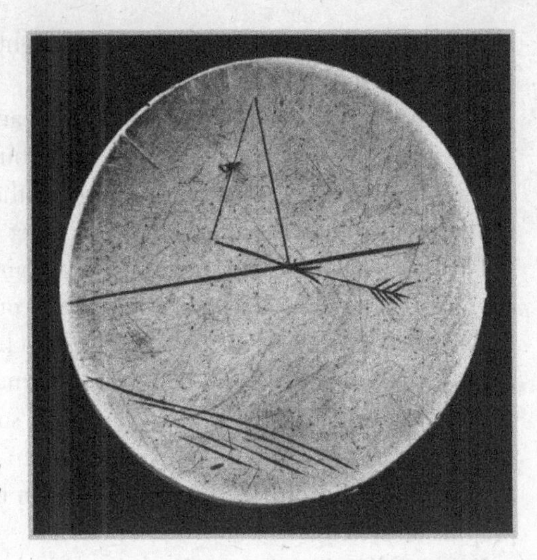

On the other side, Cayley properly identified the physical forces governing flight.

other early dreamers envisioned flat panels as wings, Cayley realized that curved or cambered surfaces lift better than do flat ones. He thus invented the concept of the airfoil, as the wing's aerodynamic profile is called (an airfoil is the shape you would see if you sliced vertically through the wing parallel to the fuselage). Cayley was also the first to realize that dihedral, an upward angle to the wings as they extend outward from the fuselage, increases lateral (side-to-side) stability by making the airplane's wings self-righting.

In 1804, Cayley built himself a *whirling arm,* a device used by ballistics researchers to study the flight characteristics of cannonballs and other military projectiles. Putting this device to a new use, Cayley investigated aerodynamic lift and drag and evaluated different airfoil configurations, backing up these painstaking experiments with rigorous mathematical calculations. This was the first time in history that scientific tools and methods were applied to the investigation of flight, making Cayley the world's first aeronautical engineer.

That same year, he built a model to the configuration he had defined five years earlier. About 5 feet (1.5 meters) long, this first airplane glider consisted of a fore-and-aft pole with a kite-like wing mounted near the front. This wing he inclined at an angle of 6 degrees. At the rear, Cayley had a cruciform tail angled downward at 11.5 degrees to offset the upward-angled wing's tendency to pull the glider into a loop. This served to give the glider longitudinal (fore-and-aft) stability.

Cayley attached this cruciform tail with a universal joint. Between flights, he could loosen it and alter its positioning relative to the fuselage. He would then fly the glider again to observe the results. Just as cleverly, this model included a weight that Cayley could shift forward or aft on the fuselage to study the effects of shifts to the center of gravity.

His handiwork pleased him greatly. "It was very pretty to see [this model] sail down a steep hill," he wrote with satisfaction, "and it gave the idea that a larger instrument would be a better and safer conveyance down the Alps than even the surefooted mule, let him mediate his track ever so intensely. The least inclination of the tail towards the right or left made it shape its course like a ship by the rudder."[4]

To again put things in historical perspective, 1804 was when the world's first rail-mounted, self-propelled steam engine—the locomotive—was tested in Wales. Railroads didn't even exist yet, and here was Cayley thinking of air travel.

Five years later, Sir George Cayley shared his findings and beliefs in "On Aerial Navigation," a landmark three-part treatise published in 1809–10 by Nicholson's *Journal of Natural Philosophy, Chemistry and the Arts*. "I feel perfectly confident," he asserted in this paper's first installment, "that this noble art will soon be brought home to man's general convenience, and that we shall be able to transport ourselves and families, and their goods and chattels, more securely by air than by water."[5]

Also in 1809, Cayley constructed an almost full-size glider to his formula. It had a fabric wing 300 square feet (28 square meters) in area. Running with this glider into a breeze, he found it lifted him so strongly his feet lost traction on the grass. Occasionally it even plucked him briefly off the ground and he soared through the air.

A polymath, Cayley pursued interests in optics, theater design, prosthetics, ballistics, electricity, heat engines, and land reclamation, making contributions to all these fields. A prolific innovator concerned for human welfare, he invented a self-righting lifeboat. Pondering history's first train accident, he came up with the cowcatcher, seatbelts, and automatic signals for railway crossings. And long before military tanks or construction equipment needed caterpillar treads, he patented one he termed the *universal railway*.

Cayley even dabbled in physics. Reading Sir Isaac Newton, who died a half century before he himself was born, he took his famous predecessor to task for having too simplistic a theory of lift, a failing he corrected.

In addition to these pursuits, Sir George helped found and chaired a polytechnic institution in London, and for a period he represented Scarborough as a member of Parliament. He belonged to the progressive Whig party, naturally.

With such duties and diversions, it was only late in his life that Cayley found time to return to his first love. Progressing to full-scale flight tests, he constructed a small glider along the lines of his 1799 design except that this one had three fabric wings stacked vertically. So well did this triplane glider perform with ballast that Cayley allowed a ten-year-old boy to make at least one untethered hop in it. Here in 1849, then, was history's first free flight of a manned heavier-than-air vehicle, albeit an unpowered one.

In 1853, at age seventy-nine, Cayley completed a large single-wing glider and induced one of his servants, a coachman, to fly it for him. Much more ambitious than the previous experiment, this flight took place at Brompton Dale with Cayley's family, friends, and servants looking on.

The coachman was more comfortable with carriages than gliders. He settled apprehensively into its seat and nervously grasped a tiller attached to rear control surfaces while his master communicated final flying instructions. Collective help was brought to bear, and the machine was sent rolling down a steep hill. As the glider gathered speed, its fabric wing billowed and lifted it into the air. Witnesses saw it dip alarmingly and then level out in time for a jolting touchdown some 900 feet (275 meters) from the takeoff point.

Cayley was among those who rushed up as the coachman climbed out. "Please, Sir George," the servant blurted, "I wish to give notice—I was hired to drive, not fly!"[6]

The flight at Brompton Dale took place exactly fifty years before two brothers in the United States would invent what this nobleman had conceived of in England. Cayley's contributions to the Wright brothers' coming success are too numerous to count even though they are largely indirect. For example, normal wheels would have been too heavy for gliders, so Cayley invented a tension-spoke wheel using taut strings to keep its hub centered. Later versions of this lightweight wheel featuring metal spokes would prove critical to the bicycle, an invention that helped set those two American brothers on the path to success.

Sir George Cayley died in 1857, at age eighty-three. Ultimately, the most astonishing thing about him is how much he accomplished with no one else's thinking to build upon. Two centuries later, this North Yorkshire genius remains early flight's towering intellect, a claim rivaled only by Wilbur Wright.

Cayley's unshakable belief in the airplane and his scientifically based description of it encouraged others to share the vision of flight. Chief among his disciples was a young man of action in England's southwest. Born in 1812, William Samuel Henson worked as an engineer in the lace industry in Somerset. Just how Cayley's writings came to his attention is not known, but they changed his life. Above all, Henson was thrilled by the Yorkshire baronet's conviction that heavier-than-air flight was not only possible but imminent.

William S. Henson, Cayley's enthusiastic disciple, set out to create a working airliner before the middle of the nineteenth century.

Bursting with ideas and enthusiasm, Henson decided to get a jump on the world by designing its first airplane. As if that weren't enough, he further determined that it should be a practical transport machine that people could put into immediate commercial use, not merely a research prototype, as one might reasonably have expected.

This boundless confidence was perhaps a reflection of the heady times in which Henson lived. The industrial revolution was under way, bringing newfound might and muscle to Britain's ever-expanding empire, already the largest the world had ever seen. Distant lands were being explored, science was broadening man's horizons, and newspapers announced exciting new discoveries almost daily. Anything was possible, or so it seemed to Henson.

He threw himself into the task of creating his airplane. The first order of business was to draw up construction plans for the machine, which he christened *Ariel*. That name didn't stick, and the world knows it as the Henson Aerial Steam Carriage. Although not destined to be built and incapable of flight if it had been, it was nevertheless such a rich imagining of manned flight that it would have a seismic impact on the quest for wings.

There were many innovations in Henson's design, which featured a rigid cambered wing, a streamlined enclosed fuselage, a three-wheel landing gear, and a bird-like tail complete with stabilizers, elevator, and rudder. An internally housed steam engine turned two pusher paddle propellers with six blades each. Bracing wires helped hold everything together.

Finishing up his drafting, Henson applied to the British government for a patent to protect his "invention," as he termed it. His unshakable certainty that it would fly perhaps being contagious, this patent was granted around Henson's thirtieth birthday. The following year he incorporated with three friends under the name Aerial Transit Company. Here, then, was history's first aviation firm—in 1843.

What's so fascinating is that Henson set for himself the huge and needless additional challenge of inventing not just an airplane but a working airliner. Confirming this intent, the company's patent application listed the vehicle's purpose as being "to convey letters, goods, and passengers from place to place through the air."[7]

Feeling they had a sure winner that would transform the world, the company's four officers succumbed to flight's siren lure. Besides Henson, they included John Stringfellow, a fellow engineer in the silk trade and a gifted tinkerer with carriage-building experience and expertise constructing (to risk an oxymoron) lightweight steam engines; Frederick Marriott, a journalist with a flair for public relations; and D. E. Colombine, a Regent Street solicitor who served as business manager and fund-raising director.

The company's first priority after incorporating was to raise funds for the construction of a full-size airplane. To promote their vision of the future, Frederick Marriott commissioned artwork depicting Henson Aerial Steam Carriages cruising serenely above London, India, Egypt, China, and other exotic locations. For maximum human interest, the tinted lithographs included people on the ground below reacting in wonder.

This evocative artwork captivated the public. Without realizing it, Henson and his team had made their one and only contribution to flight: introducing the world at large to a coming invention and showing people how it might be used. At least in terms of popular culture, the airplane had come into being.

Marriott and Henson together also promoted the Aerial Steam

Beginning in 1843, artwork of the Henson Aerial Steam Carriage showed the world what the airplane would someday be.

Carriage in words. The former's polish and latter's fervor are evident in this statement, read aloud in Parliament to secure the company's incorporation:

> This work, the result of years of labour and study, presents a wonderful instance of the adaptation of laws long since proved to the scientific world combined with established principles so judiciously and carefully arranged, as to produce a discovery perfect in all its parts and alike in harmony with the laws of Nature and of science.
>
> The Invention has been subjected to several tests and examinations and the results are most satisfactory, so much so that nothing but the completion of the undertaking is required to determine its practical operation, which being once established, its utility is undoubted as it would be a necessary possession of every empire and, it were hardly too much to say, of every individual of competent means in the civilised world.[8]

Ignoring the inevitable laughter and derision, they set to work to show the world. Unfortunately, however, reality soon intruded on their unbridled enthusiasm. A small model of the Aerial Steam Carriage briefly hopped in 1844, but two larger models—the second with a wingspan of 20 feet (6 meters)—showed no signs whatsoever of flying. By then the press and public were actively questioning Henson's sincerity. Amid the growing hue and cry, Henson found his company denounced as a hoax to defraud investors.

By 1848, a heartsick William Henson saw that it was pointless to persevere. Abandoning his dreams, he dissolved the company and immigrated with his wife to the United States the following year. Starting over as a machinist and civil engineer in New Jersey, he raised a family and lived four more decades, never again to indulge his passion for flight. Frederick Marriott likewise relocated to America, settling in California to become a respected newspaper publisher. Stringfellow remained in England and found limited success dabbling with steam-powered models.

Over the succeeding decades, all three men saw frequent reminders of their dashed hopes, particularly after 1880, when the invention of halftone reproduction made photographs and other illustrations commonplace in newspapers, magazines, and books. For more than a half century Marriott's evocative lithographs were the staple of publishers needing to illustrate anything with a futuristic theme.

Although the Henson Aerial Steam Carriage was just "such stuff as dreams are made on," its cultural impact was astonishing because it crystallized in people's minds the *idea* of the airplane. All the proper elements—wings, fuselage, tail, landing gear, and propulsion—were there in more or less the right place thanks to Henson's imaginative elaboration of Cayley's scientific insights.

Now all that remained was to invent the real thing.

2 BIRTH

WILBUR, ORVILLE, AND THE WORLD

> Learning the secret of flight from a bird was a good deal like learning the secret of magic from a magician. After you once know the trick and know what to look for, you see things that you did not notice when you did not know exactly what to look for.
>
> —ORVILLE WRIGHT (1871–1948)[1]

The morning dawned raw and cold on North Carolina's Outer Banks. Gulls wheeling high in the sky cried a timeless greeting to the surf. Nothing ever changed, and yet everything was about to. It was December 17, 1903.

A stiff wind redolent with salt and sea grass scoured a landscape hewn by the elements. Wintry sunshine caught round-shouldered sand dunes, bathing them in its glow and glinting where puddled rainwater had frozen solid during the previous night's frost.

Seven figures and an odd contraption shared a vast open area between the dunes and the Atlantic Ocean. Flat sand ran for a mile or two in every direction except back to big Kill Devil Hill and its smaller companions, which were a quarter mile away. There was nothing to break the wind.

Three of the figures were able-bodied helpers recruited from the nearby Kill Devil Lifesaving Station: John T. Daniels, W. S. Dough, and A. D. Etheridge. Two were locals braving the wind out of interest, W. C. Brinkley and a boy named Johnny Moore. A general invitation had been extended, but nobody else had come.

The two central figures, Wilbur and Orville Wright, huddled around the contraption making adjustments. Just then a great blue heron skimmed low across the sand. They broke from their labors to watch the unhurried flap of its wings.

In their thirties, these brothers from Ohio had been coming to this remote stretch of Atlantic coastline for several years. Located near the fishing village of Kitty Hawk, North Carolina, it offered what they needed for their experiments: open space, cushioning sand, and constant wind.

This morning, late in the season, there was if anything too much wind. They waited for it to abate so they could test the open-frame flying machine that was the culmination of their efforts. Fashioned of

Orville and Wilbur Wright with King Edward VII of Great Britain.

wood, wire, metal fittings, and unbleached muslin fabric, the Wright 1903 Flyer represented their best conception of what a heavier-than-air flying machine—an *aeroplane*—should be. Years of learning and experimentation performed under the rigorous discipline of scientific methodology had convinced the brothers that it combined the technological elements needed for human beings to fly.

To fly! This intoxicating wish—the headiest of ambitions—seemed a pursuit fit more for gods than mortals. But for those inspired by its fine madness, as were the Wright brothers, even the slimmest chance of success was worth the dual risks of injury and opprobrium.

As the twentieth century began, the latter was a grave risk. Most people wanted nothing to do with heavier-than-air flight and ridiculed those who did. Man-carrying flying machines were the realm of naive dreamers, self-deluded crackpots, and fantasy writers such as Jules Verne and H. G. Wells.

But birds flew, Orville Wright reminded himself, his eyes on the heron as it disappeared into the distance. Insects and bats also flew, and so did gliders such as the ones he and Wilbur had tested during their previous stays at Kitty Hawk. Of course, gliding wasn't flying. To claim the latter, you had to sustain yourself in the air, not simply descend safely to the ground.

The Flyer had first been ready three days before. The brothers had tossed a coin to see who would have the honor of trying it first. Wilbur had won, but on takeoff he pitched the nose too high. The craft had stalled and fallen back onto the sand. Now the damage had been repaired and it was Orville's turn, but the wind would not moderate.

With some trepidation the brothers elected to try it anyway. Their calculations told them that the prevailing conditions were within their machine's operating capabilities. Later in life, Orville would look back in astonishment at this decision. Virtually all early aviators waited for calm conditions before daring to fly, and they had flown in a ripping wind.

Orville solemnly shook Wilbur's hand, his heart beating with the same excitement that animated his brother's face. Sand and salt stung his eyes as he slid onto the Flyer's lower wing. He settled his hips in a wooden cradle that was part of its control system and gripped the two wooden sticks before him.

Wilbur and John Daniels attached a coil to the 12-hp engine mounted on the wing beside Orville. Connected to dry-cell batteries not carried aboard the plane, the coil provided electricity for starting. Sand scrunched beneath their shoes as the men positioned themselves before the craft's aft-facing propellers. On Wilbur's command, they yanked the propellers, and the Flyer started up with a clatter. Wilbur dashed around and unclipped the coil.

Orville moved the lever that released a restraining cable and simultaneously started automatic recording instruments aboard the Flyer. The winged craft accelerated down its wooden guide rail, Wilbur running alongside to steady its wings.

Mindful of what had happened before, Orville gently nudged the Flyer aloft, careful not to hold too much up-elevator. Pitch control was extremely ticklish, he realized. Despite his best efforts, the craft porpoised through the air.

But he was flying! A full 120 feet (37 meters) of sand passed beneath the Flyer before its skids contacted the ground after twelve seconds.

History had been made—the world's first airplane had flown.

Wilbur watches as Orville performs history's first airplane flight on December 17, 1903.

To those watching, it had unfolded with uncanny slowness. The Flyer, designed to cruise at about 32 mph (50 km/h), had flown into a wind that averaged 25 mph (40 km/h). Like a fish swimming against a fast-moving current, its progress over the sand had been reduced to 7 mph (11 km/h)—slow enough that Wilbur might have kept up had there been firm ground to run on.

As instructed, Daniels had scrambled to Orville's pre-positioned box camera. By the time he took his famous picture, Wilbur had given up the chase and was watching in awe. His awkward body posture in this iconic image—a treasured instant of human achievement captured on glass plate—speaks to wonder and triumph as few photographs ever have.

Taking turns, the brothers flew three more times that day. All takeoffs were performed from level ground solely by the power of the engine that they had designed and built. Those flights averaged about 10 feet (3 meters) in altitude. The longest—the Flyer's fourth and Wilbur's second—lasted a full fifty-nine seconds and covered 852 feet (260 meters) over the ground, or more than half a mile through the moving air.

Elated, the team carried the craft back to camp and set it down in what they thought was a safe spot. An exhilarated conversation ensued that was cut short by a savage gust of wind. It lifted the Flyer and began flipping it over.

"All made a dash to stop it but we were too late," Orville later wrote. "Mr. Daniels, a giant in stature and strength, was lifted off his feet and, falling inside between the surfaces, was shaken about like a rattle in a box as the machine rolled over and over. He finally fell out upon the sand with nothing worse than painful bruises."[2]

The world's first airplane would never fly again, but it had fulfilled its purpose. A quarter century later, its disassembled remains—which survived a flood during storage in the Wright basement—were lovingly restored by Orville, the surviving Wright brother.

Today the Wright 1903 Flyer, one of the crown jewels of the Smithsonian collection, is displayed for all to see at the National Air and Space Museum, in Washington, D.C. Every human being can feel proud of this world heritage artifact.

Viewing this proto-airplane, now more than a century old, one might well wonder how two brothers from Ohio came to solve the horrendous unknowns of flight. How did they conjure into being, whole and working, something that had never before existed on the face of the earth?

Traditional explanations have it that the Wrights succeeded through sheer pluck, determination, and good old American know-how. This unhelpful answer, the product of mythmaking by a proud nation, provides no insights. Worse still, it falsely implies that the airplane could not have been invented anywhere but in the United States.

Then how did the Wrights change the world on the sands of Kitty Hawk? The answer has two parts, the first being the immediate human tale as told in this chapter. The second, the technology side of the story, is sprinkled among this book's subsequent chapters, each of which examines a different aspect of flight technology.

The Wright brothers were self-taught aeronautical engineers with little if any formal scientific training. Gifted and intelligent as they were, neither Wilbur nor Orville actually graduated from high school because circumstances conspired to prevent them from receiving the diplomas they deserved. So how did these high school dropouts (technically, at least) succeed?

The traditional view of the Wrights is of sober midwesterners who worked in self-imposed isolation far from inquiring eyes. Being sons of a Protestant bishop who believed showmanship and public spectacle were unseemly, they shunned publicity and instead focused on the task at hand.

While there is some truth to the above, the actual story is far richer.

Born in 1867 and 1871, respectively, Wilbur and Orville grew up in Dayton, Ohio, amid a close-knit family that included two older brothers and a younger sister. Between them, there had also been a set of twins who died in infancy.

Young Wilbur and Orville shared interests. Although very different, they sought each other out and could usually be found in each other's company. They played, explored, and discussed their world with a closeness rare among siblings.

These discussions could become heated—the boys called it "scrapping"—when they found themselves arguing different sides of an issue or idea. The intellectual rigor they summoned to make their respective cases was a wellspring of creativity that contributed to their later success.

Theirs was a changing world. Industrialization had America in its throes; steam power was transforming the landscape, scientific discoveries were announced almost daily, and the far corners of the globe were being explored. Newfangled inventions such as electric lights and horseless carriages loomed intriguingly on the horizon.

For Ohioans in the aftermath of the Civil War, it was an exciting time to grow up. That shattering conflict had ended just two years before Wilbur came along. In its receding wake there arose a heady sense that anything was possible if people just applied themselves with sufficient intelligence and alacrity.

Fortunately for the boys, they grew up in a nurturing environment that encouraged playing with ideas and learning all one could about the physical world. Milton and Susan Wright both came from families with traditions of intellectual curiosity, and both were highly accomplished in their own right.

A stern patriarch, Milton Wright was also an influential writer and a social reformer. He had taken holy orders not out of religious fervor but rather because of his church's forward-looking stance on the moral and political issues of the day. Milton's pragmatic focus was on the here and now of contemporary American society, not arcane theological musings. The elder Wright had been an outspoken abolitionist until the Civil War ended the horror of slavery in North America. Believing in equal opportunity for all, he later supported women's suffrage, temperance, and other progressive causes of his day. His intellect and persuasive powers saw him rise rapidly through the church hierarchy until he became a bishop when Wilbur and Orville were still small.

Of English descent, Milton Wright grew up in Indiana when it was a frontier at the western fringe of American society. The rigors of his pioneer upbringing had imbued him with a hardy self-reliance characteristic of nineteenth-century American settlers. In turn, he instilled in his children a sense of the value of discipline, hard work, and integrity. Above all, he gave them his abiding belief that learning is the path to self-betterment.

Susan Catherine Koerner Wright played no less important a role in her children's formative years. Born in Virginia, she had loved as a small girl to linger in the workshop of her father, John Koerner, a German immigrant and master craftsman who built fine carriages. It became apparent that she had inherited his mechanical aptitude, and under his tutelage she became proficient in the use of a broad array of tools.

Relocating westward in pursuit of greater opportunity, the Koerners settled in Indiana. There Susan attended college, a rare event for American women in the nineteenth century. Excelling at math and the sciences, she also pursued a love of literature and emerged with a singularly well-rounded education.

It was at college that Susan Koerner and Milton Wright met and fell in love. Theirs was a relationship of equals that took delight in intellectual pursuits. When they married and started a family, it was no surprise that their household should boast a sizable and eclectic library. The Wright children were given free rein and encouraged to delve deeply in this library, which boasted Greek and Roman classics, European histories, biographies, scientific volumes, and the writings of naturalists. Also at their disposal were august reference works such as Ephraim Chambers' *Cyclopaedia*—now a century out of date but still fascinating—and the newer *Encyclopaedia Britannica*.

But books alone would not prepare Wilbur and Orville for the peculiar challenge they would take up as adults. Fortunately, they had also been taught another path to learning, one crucial to their invention of the airplane. This alternative source was empirical knowledge of the physical world gained through observation and experimentation.

Here Susan Wright deserves the lion's share of the credit. Clever at "adapting household tools or utensils to unexpected uses," she patiently taught her children how to build what they could imagine.[3] If they needed help with any of these creative projects, she was right there to propose how a thing might be done.

The children's maternal grandfather, John Koerner, also played a role. As he had done for his daughter, he instructed the youngsters in the use of tools. Construction techniques, rules of thumb, and reading or drawing plans became second nature to them.

In so doing, father and daughter gave the next generation a priceless gift. "We were lucky enough to grow up in an environment where there was always much encouragement to children to pursue intellectual interests; to investigate whatever aroused curiosity," Orville would later reminisce. "In a different kind of environment, our curiosity might have been nipped long before it could have borne fruit."[4]

In 1878, the family was living in Cedar Rapids, Iowa, where Milton's work had taken them. Orville and Wilbur were seven and eleven, respectively, that autumn evening when he came home with something concealed in his arms. "Before we could see what it was," Orville recounted, "he tossed it into the air. Instead of falling to the floor as we expected, it flew across the room till it struck the ceiling where it fluttered a while and finally sank to the floor. It was a little toy known to scientists as a 'hélicoptère,' but which we with sublime disregard for science at once dubbed a *bat*."[5]

Made of bamboo and cork, this clever plaything—an improved version of the Launoy and Bienvenu toy that had delighted young George Cayley almost a century earlier—employed a braided rubber band to spin two paper-covered rotors, one at each end, that turned in opposite directions. When wound up and released, it ascended energetically into the sky, delighting the boys.

The Wrights' free-flying helicopter was the invention of young Alphonse Pénaud, a Parisian confined to a wheelchair by a crippling hip disease. Yearning for lost freedom, Pénaud hit upon the idea of twisting rubber bands to power a variety of model aircraft that he conceived with delight. When he wound them up and let them go, his trapped soul soared off with them.

In addition to his helicopter, Pénaud developed two models that flew horizontally. One was a toy ornithopter that passed through the air like a rapidly flapping butterfly. The other was a fixed-wing design he called

Alphonse Pénaud.

the Planophore, an excellent flyer that drew directly from the work of George Cayley. Pénaud had stumbled across Cayley's writings and republished them in France along with aviation concepts of his own.

Pénaud was quiet, with a resigned set to his mouth and serious eyes that missed little. In his humility, he had no idea how important his efforts were or how inspirational they would be to others who shared his dreams. His ultimate wish was to someday see man-carrying flying machines. This all-consuming vision led him to design a full-scale airplane that was patently far beyond the capabilities of the day. Completed in 1876 in collaboration with his good friend and fellow flight enthusiast Paul Gauchot, Pénaud's full-scale airplane design featured moth-like wings and futuristic features such as a glazed cockpit, flight instrumentation, retractable wheels, and variable-pitch propellers. In an era of steam power, Pénaud even specified an internal-combustion engine, although that propulsion technology was only just then emerging.

Sadly, Pénaud's spirits failed him. Wearied by chronic infirmity and despondent over the growing certainty he would not see his airplane built, he placed its lovingly drawn plans in a miniature wooden coffin and delivered them to fellow aerial researcher Louis Giffard. Following this bizarre gesture, he returned home and committed suicide at age thirty.

The year was 1880. By then, two growing boys called Wilbur and Orville had worn out their first Pénaud "bat."

Casting about for fun, the young Wrights decided to build more Pénaud helicopters to the same formula. They made these next ones bigger, gleefully anticipating dramatic results. Instead they learned a lesson in physics. As Orville put it:

To our astonishment, we found that the larger the bat, the less it flew. We did not know that a machine having only twice the linear dimensions of another would require eight times the power. We finally became discouraged and returned to kite flying, a sport to which we had devoted so much attention that we were regarded as experts.[6]

Experimentation was already revealing to the Wrights immutable laws of the natural world. In this case, it was the lesson that mass is volumetric.

By way of example, consider two cubical blocks of quarried stone, one twice as large in any linear dimension as the other. If the first block is 1 meter long, wide, and high, its volume is 1 cubic meter (1 × 1 × 1). In turn, the second block, measuring 2 meters on a side, would have a volume of 8 cubic meters (2 × 2 × 2).

Yes, it would take *eight* 1-meter stone blocks to fill the same space occupied by a single 2-meter block. This explains the exponential weight increase with volume and the requirement for eight times as much power for twice the linear dimension. It is why man-carrying airplanes cannot fly with braided rubber bands, and why we need never fear the giant insects of vintage sci-fi movies: they would not be able to walk or fly.

Of course, this scale effect greatly complicated life for would-be airplane inventors.

Because Milton Wright's work required periodic moves, Wilbur Wright had been born in Millville, Indiana. After he came along, their father relocated the family to Dayton, where Orville was born. A growing urban center at the confluence of several rivers in southwestern Ohio, Dayton was a picturesque cultural crossroads that offered a wonderful quality of life. Although Milton's postings saw the family move again from time to time, they considered Dayton their home and returned there once and for all in 1884.

"If I were giving a young man advice as to how he might succeed in life," Wilbur would say with dry humor, "I would say to him, pick out a good father and mother, and begin life in Ohio."[7]

It was indeed a good life in the Wrights' big frame house. "No family ever had a happier childhood than ours," Katharine Wright, the youngest sibling, later said. "I was always in a hurry to get home after I had been away half a day."[8] Inside the tight-knit group, Wilbur was called Will, Orville was Orv, and Katharine Kate.

To this day the Wright brothers remain Dayton's most celebrated sons. Of the two, Wilbur was the more cerebral and introspective. Tall, lean, and blue-eyed, he looked ascetic but was in fact vigorously athletic. With his clear intellect and exceptional memory, he stood out as the most accomplished of the five siblings.

Orville was the paradox. Shy with outsiders, he was outgoing and a bit of a prankster within the family group. He also possessed the greatest flair for tinkering with things. Fascinated by technology, he opened up gadgets to see how they worked and fixed them if they didn't.

It came as no surprise to the family when, while still in his teens, Orville built himself a working printing press. Its remarkable design, which could turn out a thousand printed sheets per hour, drew on an odd assortment of salvaged parts, including a tombstone slab, the fold-down top of a horse-drawn buggy, and discarded firewood.

Hearing of it, a professional printer stopped by to see it in action. He examined the machine from all sides, reportedly even crawling beneath it. "Well," he announced in consternation, "it works, but I certainly don't see how."[9]

Wilbur's mechanical skills nearly rivaled those of his younger brother, but cerebral pursuits were more likely to claim his attention. Studies had been planned for him at Yale in Connecticut, and he seemed destined for an academic life. That changed with a sports injury suffered at age nineteen in a hockey-type ice game called shinny.

It was the winter of 1885–86. A shinny bat flew from a friend's hand and caught Wilbur full in the face. Although he lost some teeth, he initially seemed to recover from the accident. Then heart palpitations and other complications—perhaps including depression—assailed him. Wilbur withdrew into the Wright home and found solace in books. Other family members worried about him, as he appeared to give up all plans for formal studies outside of the home. Yale was forgotten.

Things were already difficult for the family because Susan Wright had fallen ill with tuberculosis. The older Wright boys were now grown and gone, Milton was away on church business, and Orville and Katharine were in school. Consequently, the burden of caring for his failing mother fell primarily to Wilbur.

They were much alike and he was devoted to her, carrying her up the stairs of the family's house in the evening and down again the following morning. Sadly, Susan—who had done so much for her sons—would not live to see them invent the airplane. She died in her late fifties in 1889. Young Katharine, then fifteen, pitched in and helped pull the family through the ordeal by taking over her mother's duties.

Orville, meantime, had been doing his best to keep his brother's spirits up. He worked summers as a printer's assistant to learn the trade. In the mid-1880s, he talked an apathetic Wilbur into joining him in a specialty printing firm in Dayton they called Wright & Wright. This small but flourishing business marked their first successful collaboration.

Of all things, a new craze sweeping the United States is what finally pulled Wilbur out of his doldrums. Bicycling was all the rage in America thanks to the 1887 introduction of an innovation from England called the safety bicycle. The bicycle had been around for many years, of course, but the previous incarnation—the ordinary—featured a very tall front wheel that made it unwieldy to mount and control. In contrast, the new British design used wheels of equal size. It also introduced pneumatic tires that eliminated jarring vibrations. Being easier to mount and ride, it opened to the masses the freedoms and joys of cross-country cycling.

The brothers purchased bikes of their own starting in the spring of 1892. Orville's was a new machine, Wilbur's used. Indicative of their different personalities, Orville, then twenty, loved to race flat out across short distances—he proclaimed himself a "scorcher"—whereas Wilbur paced himself in therapeutic tours through town or its verdant Ohio surroundings.

Thanks to their local reputations as masters of things mechanical, they found themselves repeatedly approached by friends wanting them to adjust or fix their bikes. This clamor and their love of the sport led them to open a small bicycle rental and repair business. Perhaps inevitably, they soon also began thinking of how to improve on this invention. By 1893, their creative talents were focused squarely on the design, manufacture, and marketing of bicycles of their own design.

Dayton was a great place for this latest business venture. The small city offered the foundries, machine shops, and light industry needed to support entrepreneurial manufacturing operations such as the Wright Bicycle Company. Being a commercial crossroads, Dayton also provided ready access to stock items from elsewhere.

Whatever the Wrights needed to build bicycles—metalworking, castings, bicycle chains, sprocket gears, or rubber tires—they learned to make themselves, have others manufacture locally, or procure nationally. It was invaluable experience for the challenges ahead.

Now successful business partners in their twenties, the brothers spent most evenings, weather permitting, on the verandah of their house. Orville sat upright, arms folded. Wilbur slouched back on his shoulders as they toyed with ideas and discussed events. Neither would ever marry, and money meant little to them. They shared a single bank account, each drawing according to need, neither caring what the other took.

In 1895, the discussion turned to the exploits of Otto Lilienthal, Germany's renowned aerial pioneer. An article showed captivating images of Lilienthal flying in wings that he had fashioned himself. Interested as they were in the German's experiments, neither brother had any inkling they might themselves soon take up flight's challenges.

Otto Lilienthal.

That changed one hot August day the following year when word arrived of Lilienthal's death from injuries sustained in a glider crash. Wilbur came across the sad news and waited till later

Lilienthal in flight, 1894.

to read the newspaper account to Orville, who lay severely stricken and close to death from typhoid fever (this same disease would claim Wilbur in 1912, at age forty-five).

After Orville recovered, they pored over images of the bearded German experimenter wearing his ingenious wings. Some depicted him standing atop a hill like an athlete waiting to race forward. Others showed him soaring fearlessly through the air, legs splayed like the talons of a bird.

The obituaries suggested that, like Icarus of the ancient Greek myth, Germany's flying man had simply dared too much. For his audacity he paid the ultimate price. If so, words uttered before his death showed he felt flying was a cause worth dying for.

"*Opfer müssen gebracht warden,*" Lilienthal had said.[10] Sacrifices must be made.

Wilbur couldn't get the event out of his mind. It started him thinking all over again about the activity then under way to solve the "problem of flight," as it was called. He and Orville followed it in the papers.

Only a few months before, for example, Samuel Pierpont Langley, the American physicist and astronomer at the helm of the Smithsonian Institution, had launched a model airplane off a houseboat on the Potomac River south of Washington, D.C. The Aerodrome No. 5, as he named it, weighed 25 pounds (11.25 kilograms) and measured more than 13 feet (4 meters) in length and wingspan. Powered by a 1-hp steam engine, this craft flew about two-thirds of a mile (1 kilometer) in wide circles before settling into the water.

Samuel Pierpont Langley.

Langley actually misnamed his craft an *aerodrome* in an attempt to call it "aerial runner" in Greek. Since the suffix *-drome* draws from *dromos*, meaning "racecourse" or "field," the word *aerodrome* would soon come to denote a flying field, not a flying machine. This etymological error notwithstanding, the Aerodrome No. 5 had demonstrated history's first extended flight by a manmade, powered, heavier-than-air device of significant size and weight.

Langley Aerodrome No. 5, 1896.

Could Langley's Aerodrome be scaled up sufficiently to carry a human being? He certainly thought so. Buoyed by his success, he worked to that end in an initially secret program that received U.S. government support.

Even more recently, Octave Chanute—America's most celebrated railroad engineer—had convened a group of aerial experimenters at the shore of Lake Michigan in Indiana, just one state over from Ohio, to test manned gliders of varying designs. Like Langley, Chanute was elderly and did not himself attempt to fly.

It helped that, despite broad derision, men of their obvious qualifications were willing to openly explore the possibility of human flight. But Wilbur didn't need any convincing on that score. "For some years," he later wrote of his fascination, "I have been afflicted with the belief that flight is possible to man."[11]

Lilienthal's death brought this to the fore. Flying became a subject of conversation, thinking, and daydreams for the brothers. They told themselves their interest was purely for the sport of gliding, but there was more to it than that. Supposing for an instant, then, that a machine could indeed be built that people could navigate through the clouds, how did one set about designing such a thing?

On May 30, 1899, Wilbur took pen in hand and wrote a letter to the Smithsonian Institution. "Dear Sirs," begins this famous missive on Wright Bicycle Company letterhead, "I have been interested in the problem of mechanical and human flight ever since as a boy I constructed a number of bats of various sizes after the style of Cayley's and Pénaud's machines. My observations since have only convinced me more firmly that human flight is possible and practicable."[12]

In this and other ways, the brothers gathered what little published information was to be had. Even with foreign works helpfully translated by the Smithsonian, there was little to work with. They pored over the meager body of research.

Fortunately, additional help and encouragement came through a correspondence Wilbur had entered into with Octave Chanute. Now retired, the eminent U.S. civil engineer was traveling widely in a self-appointed role as clearinghouse for information about flight experimentation around the world.

The Wrights kept a low profile in their research. The reason initially was their focused work ethic combined with a natural distaste for immodest or unseemly displays. Later, as they progressed, there also arose a natural desire not to give away too much until a key patent they had applied for was granted.

Octave Chanute.

A century later, this self-imposed isolation remains a source of misunderstanding about the Wrights. Viewed superficially, it suggests that an absence of outside ideas, influences, and distractions somehow allowed them to usher the airplane into being. In fact, nothing could be further from the truth; they succeeded *because* of other people's ideas, and those ideas came from around the world.

Ask aviation buffs to name the world's first multinational airplane program and chances are many will name the Anglo-French Concorde supersonic transport of the 1960s. If so, they're off by six decades, because that accolade belongs to history's first airplane.

In addition to its U.S. heritage, the Wright 1903 Flyer also boasts an Australian, Belgian, Dutch, English, French, German, and Swiss pedigree. These nations had a direct and immediate hand (sometimes more than one) in the Wrights' supposedly solitary success at Kitty Hawk.

For example, Dutch-born Swiss scientist Daniel Bernoulli in the 1700s first identified the relationship between pressure and velocity in fluid flows, helping to explain aerodynamic lift. A British-born Australian, Lawrence Hargrave, came up with the box kite in 1893—a key invention, as we shall see. Belgian-born French inventor Jean-Joseph-Étienne Lenoir gave the world its first practical internal combustion engine in 1859, while Germans such as Nikolaus August Otto and Gottlieb Daimler quickly improved on it. And still other countries can claim more tenuous connections through contributors who had no inkling their discoveries would someday be applied to the field of human flight.

Consequently, there is little if anything uniquely American about the Wright Flyer or its development. And much as the Wrights loved their hometown, the same goes for Dayton because many other places in the world offered a similar combination of intellectual openness and supportive light industry. In fact, the airplane might just as well have been invented in Manchester, Munich, Perth, Rio de Janeiro, or Toronto.

But it wasn't. It was invented in the United States. The reason was the bicycle.

The invention of the airplane was a battleground for two warring paradigms about what the airplane would be like. Paradigms are mind-sets created by what we think we know. Depending on how closely they match the actuality, these mental models either can help us succeed or can place blinders over our eyes that keep us from perceiving what we later realize was obvious all along.

Working under the right paradigm helped Orville and Wilbur to succeed even as a wrong one sabotaged the hopes of Europe's many experimenters. But for this situation, the French—who felt they had invented flight because of the success of the Montgolfiers in 1783—might well have been first. If so, the airplane, like the automobile before it, would have been a European invention.

What led Europe's aerial experimenters down the garden path? Ironically, it was William Samuel Henson, Cayley's eager disciple. Or more accurately, it was the powerful sway of Henson's persuasive vision of what aviation would be.

First published in the early 1840s, the engraved illustrations of the Henson Aerial Steam Carriage continued to appear off and on in newspapers, magazines, and books for more than a half century. More thrilling artwork of heavier-than-air flying machines was hard to imagine, and the very sight of this aerial stagecoach spurred Europe's aerial experimenters to redouble their efforts. Unfortunately, however, it also handed them a lot of incorrect notions.

The concept of an aerial carriage brought with it a concomitant expectation that people would drive airplanes around the sky making flat turns as they did in horse-drawn vehicles. This unquestioned assumption shaped how France's early experimenters approached airplane design, and it cost them dearly.

Part of Henson's paradigm worked. For example, airplanes would indeed pitch their noses up or down to climb or descend. This was intuitive because horse-drawn carriages do just that when traversing hilly countryside. But carriages don't tilt sideways, or at least not very far, because that leads to a catastrophic upset.

Henson's vision told Europe's early experimenters that their airplanes must not be permitted to tilt side to side or else catastrophe would ensue. To ensure that this never happened, some experimenters used strongly upward-angled wings (Cayley's idea of dihedral) so that the airplane would be self-righting in flight. Others placed vertical fore-and-aft fabric panels between the wings of their biplanes to prevent sideslips. Both these features suggest that Europe's pioneers were terrified of *banking*, or dropping a wing in flight.

Another place where Henson's Aerial Steam Carriage paradigm misled people was the vital issue of controllability. Controlling horse-drawn vehicles does not require constant active involvement on the driver's part. The horses are set in motion and the reins are not used again until the horses needed further instruction, whether it is to speed up, slow down, change direction, or stop.

Consequently, Europe's "early birds" were remarkably cavalier about controllability. To them, all one needed to do was create an inherently stable craft whose wings never dropped to either side. After nosing this vehicle aloft, one would simply "drive" it around the sky.

A wealthy Brazilian named Alberto Santos-Dumont performed Europe's first heavier-than-air flights late in 1906. His 14-*bis*, a marginal aircraft, was largely uncontrollable, but that didn't bother him; his goal was simply to get into the air. This disregard for a key requirement of flight was then so pervasive that more than a year would pass before any European figured out how to actually land where he had taken off.

Wilbur and Orville worked under a different mind-set. They too had seen Henson's artwork, but it didn't sing to them because they were bicyclists. Their intimate association with this vehicle, its operation, and its manufacture led them to approach flight development in a different way than their European counterparts.

Wilbur and Orville were not in the least scared of tilting to one side or the other in flight. Banking in flight seemed natural to them because a bicyclist leans into turns. What's more, they understood from the outset that the airplane needed to be controllable around all three axes and that the pilot had to be intimately involved with this process while aloft. These two insights were intuitive because the bicyclist must constantly direct his two-wheeled vehicle by means of a combination of active balance and coordinated use of handlebars, acceleration, and braking. If the bicyclist doesn't stay on top of these things every minute, he's in for a spill.

Today everybody understands that an airplane must tilt to turn. It seems funny to think anyone ever thought otherwise. But on the eve of the twentieth century, people simply did not know. Although bird flight certainly suggested the truth, wrong paradigms—such as Henson's idée fixe of an aerial carriage—have tremendous power to blind people to the obvious.

3 CONFIGURATION

SHAPES AND IDEAS

There are in fact two things, science and opinion; the former begets knowledge, the latter ignorance.

—HIPPOCRATES (460–377 BCE)

Configuration was the first great challenge for aviation's inventors. After all, they couldn't very well build their flying machines until they had decided how to lay them out.

But what form should the airplane's fuselage, or body, take? And where along this fuselage should the wings, engine, control surfaces, cockpit, and landing gear be mounted? How many wings should there be? If more than one on each side, should these lifting surfaces be mounted one above the other or in a tandem layout with one behind the next?

From our vantage point more than a century later, it seems strange that configuration was a challenge at all, let alone a vexing one. However, we benefit from hindsight. Aviation is such a part of our modern world that all of us—nervous fliers, Luddites, and small children included—pretty much know what an airplane looks like and where its parts should go.

At the beginning of the twentieth century, in contrast, there were no airplanes buzzing about overhead to provide answers. So what did people start with? For one thing, the Henson Aerial Steam Carriage; from the mid-nineteenth century onward, William Henson's artwork rode the skies of our collective imagination like Valkyries sliding on wires across the stage of a Wagnerian opera. The Aerial Steam Carriage audaciously suggested that airplanes—like ships, carriages, and trains—would soon be a useful mode of transport.

Seeing those widely reproduced illustrations prompted flight's pioneers to redouble their efforts. But even as they took heart, they dismissed the Aerial Steam Carriage's technical details as mere futuristic imaginings. Here was artistic license, they believed, not practical guidance.

That was a shame because the Aerial Steam Carriage was not entirely a flight of fancy. Backed by George Cayley's towering genius, Henson's elaborate design offered configuration help on a platter. People could have spared themselves a lot of time and grief if they had only realized it.

As 1871 drew to a close, a disastrous war launched by France on Prussia culminated at the gates of Paris. German soldiers manned fortifications, shots rang out over barricades, and cannons crumped in hostilities that would leave the City of Light more heavily damaged than it would be in both of the coming century's world wars put together.

Although the French would lose the Franco-Prussian War, their resourcefulness under fire at least gave the world its first improvised airmail service. All through the four-month Siege of Paris, balloons serenely surmounted the fray to keep the blockaded capital in contact with the outside world.

Among the massed German troops witnessing all this from below was a tall, strongly built Prussian fusilier with bright red hair. The twenty-two-year-old's upturned face betrayed wonder, a look he had reserved since early childhood for birds and anything else that flew.

Fortunately for posterity, Otto Lilienthal survived the war. Returning to Pomerania, in northeast Germany, this intense young man completed formal training and embarked on a successful career as a civil engineer. Lilienthal opened a small machine shop and fabrication plant in Berlin. In his free time, he applied his newfound engineering knowledge to the close observation of birds and experiments with airfoils. By 1878, he was using a whirling arm to test wing designs.

In 1889, Lilienthal published all that he had learned in *Vogelflug als Grundlage der Fliegerkunst* (*Bird Flight as the Basis of Aviation*), a seminal work in aviation. In it Germany's great pioneer did his best to lay out a groundwork for the emergence of human flight. Although incomplete and flawed, this volume offered those following in his footsteps a wealth of information about wings, airfoils, lift, and camber.

Like William Henson, Sir George Cayley's other disciple, Otto Lilienthal strove to build further on the Yorkshire baronet's brilliant insights. Fortunately, his success as an engineer and inventor gave him the wherewithal to undertake flight experiments, even paying for the construction of an artificially graded hill.

Picking up where Cayley had left off four decades earlier, Lilienthal became the first human being ever to make repeated gliding flights. This he accomplished with fixed wings he designed and built himself. Because Lilienthal strapped himself into these nonflapping wings and used his feet as the landing gear, these devices were by definition hang gliders. Like their modern counterparts, they were controlled in flight by shifting one's body weight.

All his life, Otto Lilienthal had dreamed of flying like a bird. At the age of thirteen, he had even fashioned a fixed-wing glider with the help of his brother Gustav. That crude effort had been entirely unsuccessful. Now, armed with superior knowledge in his early forties, he succeeded at last. Sailing through the air became his life's all-consuming passion.

This experimentation spanned five years and at least sixteen different glider designs. It won Lilienthal wide acclaim and an international following. Photographs of Germany's "flying man" were manifest proof that heavier-than-air constructions could indeed carry people aloft and that they did not have to flap their wings to fly.

To those dreaming of flight, these images were a bracing tonic. With his full beard, piercing eyes, and purposeful scowl, Lilienthal looked like a champion, and experimenters around the world took heart. But it was one flight of Lilienthal's in particular that would change the course of history.

By August 9, 1896, Otto Lilienthal had logged some two thousand flights. On that fateful summer day, the barrel-chested man stood again poised like an eagle on a crag wearing wings he had fashioned. As he had done so many times before, he raced headlong down the hill and launched himself into the air.

As he sailed aloft, a savage gust of wind slammed into him. His desperately flailing legs failed to keep his glider upright. Over he went, plummeting 50 feet (15 meters) to the ground and breaking his back on impact. Doctors, family members, and friends did their best for him, but it was no use—aviation's towering figure of the latter part of the nineteenth century succumbed the next day.

The fall that shattered Lilienthal's spine also severed the link between Cayley and those who would actually invent the airplane. At least, that's how it seems.

In 1799, Cayley had advocated designing an airplane with its wings toward the front and a stabilizing tail at the rear. Being like a bird, that much was entirely intuitive. But the Yorkshire genius went further, reasoning properly that an airplane's tail should also have vertical surfaces, and that this cruciform tail unit should be called on to contribute to both stability and control in flight.

Cayley's formula is not the only way to lay out a working airplane, as countless unconventional designs have attested over the decades. Nevertheless, the fact that his 1799 configuration has predominated from World War I to the present day shows just how presciently he identified for us the *best* configuration.

Seven years after Lilienthal's sacrifice, the Wrights succeeded at Kitty Hawk. But *how* they did it, and what their would-be competitors were doing in Europe, tells us that in this brief span, Cayley's guidance was temporarily forgotten. How do we know this? From the sheer variety of airplane layouts that experimenters were wrestling with at the start of the twentieth century. This configurational uncertainty is in fact the defining feature of flight's emergence.

The Wright 1903 Flyer has wings, an engine, propellers, and movable surfaces that deflect for control in flight. What the Kitty Hawk Flyer *doesn't* have is a fuselage. Yes, history's first airplane is all wing and no body, and that's not all that's odd about it. The Flyer also has skids instead of wheels for an undercarriage, and its elevator is at the front rather than the rear, where Cayley said to put it. Almost all airplanes since the Wrights have had rear elevators, of course.

These three configurational choices—landing skids, a forward elevator, and especially the lack of a fuselage—explain why the Kitty Hawk Flyer and its successors look strange to us today. But the Wrights had reasons for going this unconventional route.

Recognizing weight as an enemy of their efforts, they decided to identify and incorporate only those elements specifically needed for success. This they achieved through a rigorously scientific process combining observation, experimentation, quantitative analysis, and critical thinking. So unerringly did they succeed, and so highly optimized are the Flyer's technical elements, that Wilbur and Orville rank as engineering geniuses—they were true scientists, not tinkerers.

To keep weight to a minimum, the brothers selected a biplane wing structure as the heart of their airplane. Developed by their mentor, Octave Chanute, these biplane wings in turn exploited the 1893 invention in far-off Australia of the *box kite*, a new type of kite that bound parallel wings tightly together to create a rigid structure. While many kites dating back to antiquity had featured parallel wing panels, Hargrave and Chanute were the first to bring the insights of modern engineering to their construction.

In one fell swoop, this breakthrough kite technology eliminated the need for a fuselage by making the Flyer's wings—braced together with struts and wires—the airplane's skeleton. Having gone this route, the Wrights considered a fuselage superfluous and dispensed with it.

The lower wing provided space for a pilot to lie prone, and a pair of goggles sufficed for protection from the elements in a machine that would fly no faster than a horse gallops.

Had Wilbur and Orville instead followed Cayley's lead in terms of configuration, they might have come up with a more refined design with greater development potential and a longer presence on the world stage. Their airplanes might also have been easier to produce in number, a failing of the Flyers. However, the brothers would most certainly not have untangled aviation's challenges so quickly.

The configuration adopted by the Wrights has often been dismissed as lacking or backward. The brothers did fine elsewhere, critics say, but they failed to come up with a modern layout. However, this misses the point.

Imagine you're trying to solve the many puzzles of heavier-than-air flight using materials, technology, and knowledge available to you at the close of the horse-and-buggy era. Remember too that you need ready access to all parts of your prototype for adjustments, modifications, and repairs. And keep in mind that you want to observe all these components in flight to see how well they work together.

Viewed in this light, the Kitty Hawk Flyer is seen for what it is: a flying test bed. It was an airplane too, of course, but first and foremost it was a winged laboratory dedicated to the holistic solution of the problem of flight. This aerial learning platform was so well optimized to the tasks set for it that contemplating the human ingenuity behind it is truly humbling. Here is science at its best.

Remember Samuel Pierpont Langley, the scientist at the helm of the Smithsonian? Following his success with models, he oversaw the construction of a scaled-up version called the Aerodrome A. Four times as large as its predecessors, it featured two sets of upward-angled wings spanning fully 50 feet (15 meters).

Impressive as it was to look at, the Langley Aerodrome A was also fatally flawed by an aerodynamically unsound tandem-wing configuration and an excessively flimsy fuselage framework. Lacking a landing gear, the Aerodrome also had a largely ineffective control system and depended on dihedral (Cayley's idea of upward angled wings) to stay upright. The cruciform tail moved up or down for pitch control, but it functioned more as a stabilizing trim tab than an elevator. As for the rudder, Langley for some reason placed it amidships, rendering it ineffective. On the plus side, the Aerodrome A had a remarkable engine producing more than four times the power available to the Wrights in 1903.

Like Langley's models, the man-carrying Aerodrome A—piloted by Langley's assistant Charles Manly—was to be catapulted off the top of a modified houseboat on the Potomac River. The first attempt came on October 7, 1903, but the machine fell instead of flew. As one reporter put it, the Aerodrome plunged "like a handful of mortar" into the water.[1]

Disappointed, Langley blamed the launching catapult atop the houseboat, which he felt must have entangled his machine. The Aerodrome A was duly repaired and a second attempt made on December 8, 1903. Once again the machine crashed at takeoff, this time folding back on itself and breaking as it dropped into the icy river. Briefly trapped underwater in the wreckage, Manly barely succeeded in extricating himself.

Langley had received $50,000 in U.S. government support, a sum

Launched from a houseboat in October and again in December 1903, the Langley Aerodrome A twice plunged into the Potomac River instead of flying.

equivalent today to about $1.4 million. His ignominious failures put an end to the nation's public hopes of inventing the airplane. Newspapers scoffed at the idea of flight and ridiculed the waste of public funds for so frivolous a purpose.

Nine days later and some 200 miles (325 kilometers) to the south-southeast, the Wright brothers succeeded where Langley had failed. While the Wrights weren't much for publicity, they certainly had not tried to keep their achievements a secret. Nevertheless, the skepticism of a disbelieving world—undoubtedly reinforced by Langley's public humiliation—meant that it would be years before most Americans knew of their success. It didn't help that the few notices of history's first airplane flight that made it into print were wildly inaccurate.

Samuel Langley died in February 1906, at age seventy-one. His Aerodrome A survives, painstakingly restored and carefully preserved by the institution he once headed.

We have looked at two oddly configured flying machines, one successful and the other a failure. Had Langley's Aerodrome A flown, however, it would not have met the modern definition of a true airplane.

To qualify as an airplane, a machine must be manned, powered, heavier than air, capable of taking off under its own power and flying out of ground effect (not just skimming low, buoyed by a cushion of air against the ground), and controllable around all three axes. This rigorous understanding of what constitutes an airplane was largely absent in flight's early days.

Until late 1908, just the Wright 1903 Flyer and its immediate descendents qualified as airplanes. Nevertheless, confusion over this invention's definition, and the delay in understanding what had occurred in North America, led Europe to falsely claim primacy in heavier-than-air flight.

On October 23, 1906, at Paris' Bagatelle cavalry grounds, Alberto Santos-Dumont climbed into an odd-looking biplane called the 14-*bis* and opened the throttle of its 25-hp Antoinette engine. Breaking free of the ground, he sailed 160 feet (50 meters) before settling back to earth. The Fédération Aéronautique Internationale, which officially witnessed the feat, wrongly proclaimed it the world's first certified airplane flight.

Alberto Santos-Dumont.

Photographs of Santos-Dumont's machine in the air were greeted with rapturous enthusiasm. The pan-continental celebration was premature, however, because the 14-*bis* was, in fact, not an airplane. Incapable of controlled or sustained flight, this machine had biplane wings that angled strongly upward just past the propeller. Beyond the engine was a wicker basket for the pilot and a fuselage culminating in a box-like surface that tilted side to side or up and down for some measure of control.

What's funny about the 14-*bis* is that the propeller was at the rear and what looked like the tail was actually its front. Sitting on the ground, it looked like it should go in one direction when in fact Santos-Dumont intended it to go in the other. This backward layout gave aviation the term *canard* for an airplane with the main wings toward the rear, like a duck or a goose, and smaller lifting surfaces near the front.

The 14-*bis* was entirely the brainchild of Santos-Dumont, a diminu-

*The Santos-Dumont 14-*bis* performed Europe's first flight by a heavier-than-air vehicle in October 1906.*

tive man in his thirties whose limpid brown eyes, trim moustache, and air of icy aplomb won French hearts. Heir to a South American coffee fortune, Santos-Dumont had the leisure time and resources to indulge his lifelong fascination with flight. He had already made a name for himself with small dirigibles before turning to heavier-than-air experimentation in 1906.

If Alberto Santos-Dumont came up short in scientific insight, it was entirely forgivable. Nobody yet had much of that on the European side of the Atlantic Ocean. His 14-*bis*—which resembled nothing so much as a collision of Hargrave box kites—staggered into the air eight more times, the longest being a twenty-one-second wallow in ground effect.

A technological dead end, the 14-*bis* influenced no one else. Its odd configuration quickly disappeared from the scene, but not so Santos-Dumont himself. He returned to prominence in 1909 with the tiny Demoiselle, one of early aviation's most delightful successes.

Other European designs were more promising but failed to fly for one reason or another. One was by Trajan Vuia, a Hungarian-trained Romanian whose fascination with flight

Romanian Trajan Vuia with his unsuccessful monoplane of 1906.

brought him to Paris. Vuia created a Cayley-style monoplane in 1906 that, although unable to fly, presciently anticipated the modern configuration. The following year, Louis Blériot unveiled his surprisingly modern Model VII, a low-wing monoplane with a fully enclosed fuselage except for its cockpit. Although it too failed to perform, its look of evident rightness made it influential.

Even so, confusion reigned for years as to what form the airplane should take. In March 1910, another oddball flying machine made history's first successful takeoff and landing on water. Designed, built, and flown by French maritime engineer Henri Fabre on the Mediterranean coast near Marseilles, the world's first hydroaeroplane was in fact not a true airplane at all because, like the 14-*bis*, it could only perform short, straight-ahead hops.

Fabre's machine looked as if its parts were in the wrong places. One aviation historian aptly described it as resembling "an unfinished length of fence."[2] This bizarre machine had an elevator above a small wing at front with a larger wing, fixed vertical stabilizers, and a pusher engine and propeller at the rear. Its most interesting aspect was the remarkably compact set of pontoons that kept it afloat and also contributed to lift in flight.

Henri Fabre took off from the water near Marseilles and alighted again in March 1910.

Successful or not, all machines built in man's quest for wings are of historical interest because they reveal what people were thinking.

Hubert Latham strode the bluffs near Calais in mounting exasperation. His boots splashed the sodden grass. Rain poured from his greatcoat and tweed cap. From below came the crash of waves. The twenty-six-year-old aviator was known for aplomb in the air, but his nerves had frayed visibly as each new day brought more rain, wind, and fog. This dawn he could not see the English Channel below—La Manche, as the French call it—much less the white cliffs of Dover, which glowed on the horizon on clear mornings.

It was the early summer of 1909. The *Daily Mail* a London newspaper, was offering a prize of £1,000 sterling—a veritable fortune—to the first person to fly an airplane across the English Channel. Latham was determined to claim this sum and attendant fame and honors for France and the Antoinette Company.

A wealthy Parisian of British descent, Latham was tall and slender, and spoke English and German as perfectly as French. Never without his trademark accessories—a jaunty cloth cap, white cigarette holder, and wristwatch in the era when pocket watches were the norm—he traveled in aristocratic circles and described himself as a "man of the world."[3]

Above all, Latham lived for adventure. That thirst had led him to motorboat racing, big-game hunting, and aviation. Just months before daring the Channel, he had learned to fly in a graceful monoplane introduced by the Antoinette Company, which also built aero engines. A natural flier, he now served as the company's chief pilot.

The poor weather finally ended on Monday, July 19, a day that started off as bad as the rest when Latham and his team awoke at four to check the weather. But at dawn the wind suddenly abated and the damp mists began lifting. A wireless report from the British side of the Channel proclaimed ten-mile visibility.

This was all Latham needed. He mounted his Antoinette and settled into the cockpit. Starting up took twenty minutes of fiddling with the engine controls while his helpers repeatedly swung the propeller.

The Antoinette of 1909 was the world's first successful monoplane.

Finally the balky engine came to life. Latham nosed into the wind and opened the throttle, breaking free of the ground at 6:42 A.M.

The temerity of this attempt had drawn crowds to both sides of the Strait of Dover. Both shores were clogged with motorboats and yachts. Frenchmen sent Latham off with cheers and waving hats. On the English side, excited shouts broke out over news of his departure. But the mood turned to anxiety and then worry as no airplane appeared.

Instead came word that Latham had been forced down, his fate unknown. A subsequent report brought the welcome news that he was safe. His engine having failed a third of the way across the 24-mile (38-kilometer) strait, Latham had glided down to the water and skillfully ditched. The rescue boat found him perched unflappably in the half-submerged Antoinette, indulging his addiction to cigarettes.

Latham promptly ordered another airplane from the company. He would try again.

The day that Latham ditched in the channel, rival French aviator Louis Blériot announced that he too would seek the *Daily Mail* prize. Blériot and his entourage arrived at a nearby village with his Model XI, a considerably smaller and simpler machine than the Antoinette.

In his late thirties, Louis Blériot was a businessman and father of six. Stocky with sad, dark eyes and a walrus moustache, he had made a fortune manufacturing acetylene headlamps he had invented for early automobiles. Passionate about flight since 1901, he had devoted vast sums to the cause. Impatient by temperament, Louis Blériot

Louis Blériot, France's foremost early aviation pioneer.

had built airplanes to no fewer than three different configurations in 1907 and never developed or tested any of them sufficiently to get anywhere. Fortunately, he was also a quick learner and a courageous experimenter.

Blériot crashed so many times that a British journalist proclaimed him "the most daring aviator in the world."[4] This experience actually served him well, as he became a master at avoiding broken bones and other serious injuries. "I always throw myself upon one of the wings of my machine when there is a mishap," he explained, "and although this breaks the wing, it causes me to alight safely."[5]

The day before Latham ended up in the water, Blériot set a speed record at Douai in his Model XII monoplane. During that flight, an asbestos shield came loose from the airplane's exhaust pipe, resulting in third-degree burns to his left foot. It was the second time in three weeks it had happened, and the previous injury had not healed, making this new one all the more painful. But Blériot was out of funds; if he was to keep flying, he had to compete for the *Daily Mail* prize money.

Alice Blériot was her husband's devoted supporter. Squashing her fears as a wife and mother, she selflessly helped him fulfill his dreams.

In an ironic twist of fate, it was she who had made it possible for Louis to compete against Latham at Calais. While visiting friends, she had saved a young child's life by dashing to a balcony rail and snagging him just as he was about to topple off. As it turned out, the father—a wealthy Haitian planter—had an interest in aviation and wrote a check allowing Blériot to purchase the engine he needed.

Hobbling on crutches, Blériot arrived at Calais on July 21, 1909, accompanied by Alice and other helpers. His Model XII twice having injured him, Blériot instead brought his Model XI, another monoplane he had designed.

The English Channel is known for its poor weather, but this year was exceptionally bad. On July 25, however, Blériot was awakened at two-thirty in the morning with the news that stars shone in the sky. Dressing with difficulty because of his injury, Blériot arrived at the makeshift flying field where mechanics had already wheeled the small Model XI out of its tent hangar and were readying it for flight. Strapping on his helmet, Blériot took the ship up for a quick test hop. Fuel and oil were again topped off, and he set out for England. It was four-thirty and dawn was breaking.

Hubert Latham also would have flown on this date except that a friend dozed off at the wrong time and his wake-up call came late. His feelings when he heard Blériot's machine already in the air can only be imagined. By the time he too was ready, visibility had fallen and no further flying was possible.

Blériot, meantime, found himself flying in a haze over singularly uninviting waves. For ten long minutes, no land was in sight in any direction. Finally he spotted England and made a leftward course correction for wind that had blown him too far east. His intended landing field was easy to spot because of Dover Castle, one of the largest medieval fortifications in the world.

Settling over a gently sloping valley, Blériot touched down a little after five. His machine was damaged in the landing, but he emerged unscathed. Bleary-eyed and covered with oil thrown by its engine, he hopped on one foot as he released the crutches strapped to its fuselage.

Blériot conquered the English Channel on July 25, 1909.

He did not look the part of an intrepid aviator, but that didn't matter. The world went wild, just as it would with the later flights of Charles Lindbergh in 1927 and astronauts Neil Armstrong, Buzz Aldrin, and Michael Collins in 1969. Unlike those later milestones, however, Blériot's flight of 1909 is today largely forgotten, which is surprising in light of its profound social and psychological impacts.

For the first time in history, a human being had surmounted a natural obstacle by overflying it in a heavier-than-air machine. The phrase "geography is destiny" never again would be quite so true. Blériot's achievement also suggested useful applications of aviation in the future, the ultimate dream being air travel between nations.

To the British, this flight dealt the psychological blow of sudden vulnerability. For countless centuries, the English Channel had buffered the British Isles from foreign invasion. This treacherous waterway had defeated the Spanish Armada in 1588 and Napoleon's planned conquest of 1805, but now the world had changed. As alarmed Britons observed, "There are no islands anymore!"[6]

Louis Blériot himself found lasting fame as France's most significant aviation pioneer. A December 1909 crash onto a rooftop in Constantinople (today Istanbul) brought him his first broken bones. Deciding the time had come to give up flying, he instead devoted himself with great success to airplane manufacture.

As for Hubert Latham, he remained determined to conquer the Channel, although he could no longer be first. Setting out once again on the evening of July 29, he suffered a second engine failure, this time within sight of Dover, and once again had to be plucked from the water. The Antoinette firm declined his request for another airplane for a third attempt.

Latham's devil-may-care adventuring may have masked a courting of death because he suffered badly from tuberculosis, a condition worsened by his smoking. He continued taking risks until killed in 1912 at age twenty-eight by a charging buffalo while big-game hunting in Sudan.

Latham and Blériot's rivalry across the Channel served notice on the world that a dominant airplane configuration had emerged. Although the Antoinette IV and Blériot XI differ considerably in size and detail, in terms of configuration they are twins. Both are high-wing monoplanes with a similarly placed cockpit, engine and propeller, landing gear, and empennage, or tail. Both have their fuselage-mounted wing counterbalanced by a tail with horizontal and vertical surfaces that contribute to stability and control. Here was the layout George Cayley had proposed in 1799. More than a century before Blériot lifted oil-smudged goggles to contemplate England, this formula had taken flight in the hand-launched models Cayley sent sailing down his estate's grassy slopes.

Five years after Blériot's flight, the Great War broke out in Europe. It placed harsh demands on the infant technology of aviation, winnowing out what did not work well. Under this baptism by fire, George Cayley's 1799 configuration emerged victorious. There were variations on the theme, of course. During that four-year conflict, airplanes designed to the Cayley formula flew with three sets of wings (triplanes), two sets (biplanes), or a single wing (monoplanes). Among this last category were monoplanes with the wing set high on the fuselage, airplanes with the wing mounted above the fuselage (parasol monoplanes), and airplanes with the wing set low.

Virtually all airplanes built today are monoplanes, and most have low wings. Dr. Hugo Junkers, an underappreciated aviation pioneer we shall soon meet, first championed this configuration in World War I because he felt it was safer. In wartime, a low wing offers protection from ground fire and absorbs the impact of crashes. In general, of course, low wings also provide a handy place to mount the landing gear.

This configuration emerged in the United States in 1922 when the U.S. Army Air Service sponsored the creation of a racing plane called the Verville-Sperry R-3. Although a military airplane, it was named for Alfred Verville, its designer, and Lawrence Sperry, whose small company built all three examples.

The Verville-Sperry racer is astonishing not for what it did but rather for what it was. Here, shortly after World War I, was a low-wing monoplane with a wide-spaced landing gear that retracted to enclose the wheels within the underside of the fuselage. The wing itself was fully cantilevered, meaning that it required no drag-inducing external bracings.

The Verville-Sperry R-3 racer of 1922 marked the U.S. emergence of an advanced configuration pioneered by Germany's Hugo Junkers.

Except for its open cockpit, therefore, the R-3 racer embodied the configuration of the fighter planes that would fly and fight in the skies of World War II some two decades later. This airplane was so revolutionary, in fact, that the biplane era would run its course before designers caught up with Verville's thinking. Whether biplanes or monoplanes, however, it was all Cayley's basic formula.

4 FUSELAGE

OF DRUMS AND DRAGONFLIES

Balance yourself like a bird on a beam. . . .

—"COME JOSEPHINE IN MY FLYING MACHINE,"
A POPULAR SONG OF 1910[1]

Imagine you're a contemporary of the Wright brothers. Like them, you've devoted your time, energy, and intelligence to the quest for human flight. You've done your conceptual thinking and experimentation. Before you now on your workshop's plank table are sketches you've just made for the heavier-than-air flying machine you came up with.

You're sure this machine will fly—well, fairly sure anyway—if you can just figure out how to build the damn thing. But how to start?

For flight's pioneers, the logical place was the fuselage.

On the eve of the twentieth century, gasoline engines, cambered wings, a stabilizing and controlling tail, landing gear, and other foreseeable components of flight technology could all be built. The problem was that these things wouldn't carry you aloft unless you also figured out how to connect them all together.

This appeared to be the job of the body of the airplane. The fuselage would support the wings and other components, holding them in proper alignment. This was quite a challenge because aerial maneuvers or gusts of wind would put loads on these extremities like a giant's hand working a long lever.

Making a supporting frame do all this and still be light enough to fly was a devilish challenge, but that wasn't the end of it. The fuselage also has to accommodate the crew and payload (in the first airplanes, the pilot *was* the payload), and it must be aerodynamically contoured to promote the airplane's passage through the air.

What did all this add up to? The word *fuselage* itself actually gives us a clue, since it tells us what people were thinking. One of aviation's oldest terms, it derives from the French word meaning "spindle-shaped." Spindles, of course, are simple tools used by primitive cultures for weaving natural fibers into yarn or twine. Among humankind's earliest technology, the spindle—or more properly the drop spindle—takes many forms. The simplest ones are fat, smoothly rounded sticks tapering at both ends. Hobbyists use them today for the sheer tactile pleasure of making yarn the old-fashioned way.

The shape of the spindle struck at least some early flight experimenters as intuitively right for the body of an airplane. If not, why else would an obscure French term make its way into aviation's emerging lexicon? Like all linguistic hijackings, this one betrays a past mind-set.

So where did that idea come from? As always, the answer is nature. And the reason is that we human beings recognize patterns in all we see. Our brains are continuously interpreting and drawing connections.

There was a big one to draw with the drop spindle. Imagine our prehistoric ancestors fashioning and then using this tool. Its shape in their hands inevitably would have reminded them of birds or fish, the two classes of vertebrates most shaped in their evolution by the need to pass through fluid mediums with minimum resistance.

Of course, aerodynamics and hydrodynamics played no role in the design of the drop spindle. It tapers smoothly to guide fibers and not snag them. Nevertheless, spindles would by association have called to mind nature's clear lessons regarding the relationship between shape and speed.

But how did one go about engineering a flying machine along these lines? Was there an existing body of knowledge to draw on that would help us to cobble together a lightweight, structurally sound fuselage? Sadly, the answer was no.

In contrast to his bountiful insights elsewhere, Sir George Cayley's sketches and models generally showed a featureless pole for the airplane's body. He advocated streamlining, with a clarion call for "solids of least resistance," but he did not elaborate. And while he published drawings for a manned glider late in life, these fell well short of the needs of powered flight, and most researchers were unaware of them in any event.

William Henson, aviation's first popularizer, was likewise unhelpful, but for the opposite reason of providing too much detail. *An aerial steam carriage? Why, the man must be mad! One might as well ask a boat or a train to fly*, scoffed serious researchers. Visionaries dreamed of a day when aviation might achieve fully enclosed passenger cabins. However, it certainly wouldn't start out that way because they were too heavy. No, ruthless attention to weight demanded a minimalist approach to inventing the airplane.

Looking elsewhere, people turned to nature for inspiration. Predictably enough, they found it in the bird, which suggested not just the airplane's configuration but also its structure. Birds are vertebrates like us humans. We and they rely on bony internal skeletons to support our weight and carry the physical loads we encounter in life or subject ourselves to through muscular exertions. Vertebrate engineering is a triumph of evolution. Nature's invention of a structurally strong spinal column is what made large land animals possible.

Just as every arch must have its keystone, so too must a skeleton have a spine. Arranged longitudinally like the keel of a ship, it is the primary structural element in any vertebrate's body. Every other part of the body ties into this long "backbone"—actually a series of linked bones called *vertebrae*—and draws strength from it.

From aardvarks to zebras to just about every species in between of any size in the animal kingdom, they're built to this winning formula, dogs, cats, lizards, rabbits, whales, snakes, birds, and human beings included. Only in the world's oceans, with seawater to buoy their flimsy bodies, can sizable invertebrates such as the giant squid exist.

Aviation seized on this biological paradigm as the way to build an airplane (or rather its *airframe*, which is the aircraft minus its engine or engines). Like a bird, the airplane would have an internal load-bearing skeleton.

That led to more questions. First and foremost, what should this mechanical bird's bones be? The natural answer was wood, that being the preferred construction material of the nineteenth century. First-generation airplane builders would favor ash, hickory, Douglas fir, and above all Sitka spruce for their excellent strength-to-weight ratios. The Wright brothers built with ash and spruce. So did their talented American rival Glenn Curtiss, who added bamboo poles to the formula.

While every first-generation flying machine—successful or not—had a wooden skeleton, wood was far from the only material flight experimenters needed, of course. Two metals destined to play major roles in flight were also readily available.

Even then, wooden ships' hulls and rail cars were giving way to steel in a profound transformation that revolutionized global transportation on the eve of the twentieth century. Thanks to the industrial needs of this and other industries, the world had embarked on large-scale steel production. And thanks to a newly perfected method making the refining of bauxite commercially viable, aluminum too was becoming available.

As we all know, these metals—aluminum in particular—have played a crucial role in aviation. In the early days, flight's pioneers used metal sparingly in just three areas. The first was where wood was not strong or durable enough. Engine mounts, attachment points, and other key structural components and fittings were thus made of steel, as were the nuts and bolts that held them together.

The second area where metal was used was in those places that had to be fireproof, such as engine firewalls and cowlings. A *firewall* is the metal bulkhead that prevents engine-compartment fires from invading the airframe, and a *cowling* is a streamlined cover for the engine. Here aluminum filled the bill perfectly. Not as strong as steel but much lighter, it also provided brackets, clamps, and other non-structural components.

Finally, flight's inventors used metal in the form of steel wire. Used internally for diagonal cross-bracing, steel wires augmented the fuselage truss structure, significantly increasing its strength and rigidity while adding very little weight. Used externally, these "flying" wires served to rig (properly align) and brace (reinforce) the airplane's wings, horizontal and vertical tails, and landing gear relative to the fuselage.

Steel wires were thus the ligaments to the airplane's wooden bones. Often fitted with turnbuckles that allowed their tension to be changed, they were how the airplane's rigging—and thus its flight characteristics—were fine-tuned.

Of course, early airplane builders also found other uses for steel wires, such as connecting the cockpit controls to the airplane systems they actuated. But more about that later—the focus here is on structure.

Airplane builders drew many lessons from the natural world. For example, anatomical studies had revealed that bird bones are often hollow to save weight. But nature's cleverness did not end there; within these hollow bones, scientists observed cross-bracing spurs that, like reinforcing beams, offset most of the weakening that otherwise would have resulted from the deleted bone mass.

Following a similar path to weight savings, airplane builders learned to cut circles out of wood and metal in places where it would not compromise the part's overall strength. Lightening holes, internal bracing struts, and other engineering practices suggested by nature remain standard features of aerospace design to this day.

The bones and ligaments of the airplane provided its strength. Around this structure, airplane builders often applied an enveloping skin that—as is the case with birds, humans, and other vertebrates—protected but did not support. For aviation, this non-load-bearing skin generally was doped fabric.

Muslin, a cotton cloth with a fine weave, was the aviator's material of choice, although linen and other natural fabrics were also used. After being applied to the airframe, the cloth was brushed with *aircraft dope*, a varnish that draws the fabric taut as it dries to produce a hard, durable surface.

If you are ever at a small airport and you come across an old fabric-covered airplane—for example, a Piper Cub, Aeronca Champion, or Citabria—give its doped fabric a light tap with your knuckle. The thump will elicit a shivering thrum. It sounds if the entire airplane is a drumhead. Now look down this airplane's fabric fuselage or wings. You will plainly see the fuselage longerons (fore-and-aft framing members), wing ribs, and other internal bones encased by this taut skin. It's very different from the all-metal jetliners we routinely board.

The very earliest airplanes used relatively little fabric. At the slow speeds they operated, it was not required on the fuselage, where it merely added weight and hindered internal access for modifications or repair. For this reason, the earliest flying machines often appeared to the world like the dinosaur skeletons displayed by museums.

In contrast, fabric was required on the airplane's wings, tail, and control surfaces, where it guided the passage of air to provide lift, stability, and control. But even here, where it had an aerodynamic role to play, the airplane's skin never shouldered the structural loads of flight—that was strictly the job of the airplane's skeleton.

On a fundamental level, then, the bird—and by extension the human body itself—provided aviation's first successful paradigm for airplane construction. As we shall see, the insect world would later suggest a different and better way to build flying machines.

In August 1909, the world's first air meet took place at Reims, France, in a region famous for sparkling white wines. Called La Grande Semaine d'Aviation de Champagne, this exuberant weeklong aviation exhibition was a roaring success, with flight displays, speed runs, and other assorted aerial thrills.

Some two dozen flying machines from ten different manufacturers took to the air before record crowds. More airplanes still were on static display. Photographs of all this activity filled newspapers around the world. It was nothing short of a rapturous unveiling of humankind's newfound ability to travel at will through the sky.

Held in August 1909 at Reims, France, La Grande Semaine d'Aviation de Champagne was an exuberant celebration of newfound flight.

Although the Wright brothers did not attend, they were represented by others at the controls of four license-built Wright Model A Flyers. Wilbur in particular was there in spirit because it was his flight demonstrations in France the previous year that had shown others how to control an airplane in flight, making this air meet possible.

From the United States also was Glenn Curtiss, the Wrights' brash young competitor, in his open-frame biplane the *Reims Racer.* Curtiss had designed and constructed it himself right down to its engine. With no reserve airplane available in case of a crack-up, he flew it sparingly until the final day, when he threw caution to the wind.

The meet's crowning event was an all-out speed dash known as the Gordon Bennett Cup. Hubert Latham, unfazed by his recent dips in the English Channel, attempted to compete in an Antoinette but failed to make the cut. With his elimination, France's hopes rested squarely on Louis Blériot, who ultimately lost by a matter of seconds. The trophy went to Curtiss for his world record of 47 mph (75 km/h).

The French would win the following year's meet, consolidating their dominance of flight in its initial decade. This flying at Reims helped bring to a close Europe's fabled belle époque, a halcyon era of antiquarian graces, flowering arts, and galloping discovery. Four years later, the guns of August would shatter what was left of that fondly remembered time.

Glenn Curtiss wins the Gordon Bennett Cup at Reims with a blistering speed of 47 mph (75 km/h).

World War I broke out in August 1914. A conflict of stalemate and attrition, it would claim ten million lives before drawing exhaustedly to a close more than four years later. By then, Europe's borders had been redrawn and four great empires—German, Austro-Hungarian, Russian, and Ottoman—had all ceased to exist.

The Great War spurred development by calling on aviation to play meaningful roles. Airplanes barely flew in 1914, but that didn't stop intrepid aviators from strapping on leather helmets and navigating over hill and dale to gather military intelligence. Tracking troop movements and photographing enemy installations were aviation's primary wartime roles.

By 1915, the warring factions were designing airplanes for specialized uses. In addition to reconnaissance, the conflict bred fighters, bombers, maritime patrol planes, and ground-attack machines. Wartime urgency also gave rise to large multiengine airplanes. Of course, the most famous flying machines of World War I had just one engine and one seat. These were the fighters or pursuit planes (*avions de chasse*, as the French called them) that earned everlasting fame in swirling aerial dogfights.

All this activity aloft hid the war's single great gift to flight tech-

World War I brought the glamour—and danger—of early military flying into the popular culture, as this still from the 1928 movie Lilac Time *suggests.*

nology: reliable airframes. Before it all ended, human beings had taught themselves how to build truly rugged flying machines that did not break apart no matter how violently they were maneuvered. It would be another decade before reliable engines were also available.

Among the war's most rugged airplanes was the SPAD fighter series, created by France's Louis Béchereau, a graduate of the École d'Arts et Métiers d'Angers. The best of this famous lineage was the SPAD XIII, which appeared in 1917. Many of the war's most famous aviators flew this type. Top French ace René Fonck gained most of his seventy-five victories in SPADs, as did Italy's leading ace, Count Francesco Baracca (thirty-four victories) and U.S. "ace of aces" Eddie Rickenbacker (twenty-six victories).

The United States did not enter World War I until 1917, so Rickenbacker and his fellow pilots were late arrivals. Because of the backward state of the U.S. aviation industry, they had to rely on French equipment. These Americans initially flew the Nieuport 28, a structurally question-

U.S. ace of aces Eddie Rickenbacker poses with his SPAD XIII, the most successful French fighter plane of World War I.

able machine with a fatal tendency to shed its upper-wing fabric in dives. Despite these shortcomings, the Americans did well with the Nieuport.

But then they got their hands on the SPAD and never looked back. The SPAD XIII was very fast but tricky to fly and less maneuverable than the Nieuport. Still, these failings paled compared to the benefits, particularly its robust construction. If any airplane would bring you home safely, this one would.

More than just rugged, the SPAD XIII embodied remarkably advanced ideas on design safety. For example, Louis Béchereau located the main fuel tank low on the fuselage between the bottom wings. This placed it beneath the pilot's feet, more or less at the plane's center of gravity. However, Béchereau also specified leather straps to hold it in place. In the event of a fire, these straps burned through and the flaming tank fell harmlessly away, allowing the pilot—his airplane still in balance—to make a safe dead-stick landing.

Despite this Gallic cleverness, the honor of fielding the war's most technologically influential airplane goes hands down to the Central

Powers. This breakthrough German machine was the Fokker D.VII, an advanced fighter design by the Dutchman Anthony Fokker that reached the front in the spring of 1918.

The Fokker D.VII looked much like any other World War I fighter, but beneath its colorful fabric lay a skeleton of welded steel tubing, not wood. A single turnbuckle-tensioned wire looped through lugs at this fuselage's welded junctures created a braced box-girder structure for added rigidity. Atop these metal bones and beneath the fabric was curved plywood decking that rounded out the plane's lines aft of the cockpit. Significantly, the D.VII also had thicker, stronger wings, which was another German innovation. Together these structural advantages allowed D.VIIs to dispense with nearly all the bracing wires required by other World War I airplanes.

In technological terms, the D.VII looked ahead to what aviation would be between the world wars. It's no coincidence that it was the only airplane specifically mentioned in the Treaty of Versailles, the peace agreement that formally ended the war six months after hostilities ceased.[2]

Once techniques were devised for building fuselage skeletons out

Feared by Allied pilots, the Fokker D.VII was technologically the most influential airplane to emerge from World War I.

of steel, aviation quickly went this route. Steel offered greater strength than wood. It was also more uniform and thus predictable, whereas stocks of high-quality wood were sometimes hard to come by.

As airplane design progressed from art to science, this last fact became metal's greatest benefit. Unless stress loads and paths can be calculated accurately and with confidence, airframes cannot be built as light as possible yet sufficiently strong to do the job safely. Too much strength is not desirable because it means that unproductive weight is being carried, compromising performance and wasting fuel.

With the armistice on November 11, 1918, peace had returned to Europe. In marked contrast to the end of World War II a quarter century later, the Great War left Europe's industrial base largely intact. Despite the devastation near the shifting front, the factories that had forged the Central Powers' aerial weapons of war were left untouched.

With peace at hand, these factories now turned their attention to commercial aviation as a promising arena for the technical expertise they had acquired. In particular, they looked to the manufacture of airliners.

Aside from North Atlantic weather patterns, which brought miserable winter weather and extended periods of low visibility, Western Europe was a great place for airline travel to emerge. European capitals are relatively closely spaced, and except for the Alps, the continent's geography is not unduly challenging. By the start of the 1920s, there-

France's Farman Goliath airliner of 1919 put World War I technology to a new use.

Rugged and reliable, the all-metal Junkers Ju 52 trimotor was also heavy and slow.

fore, KLM and the ancestors of today's Air France, British Airways, and Lufthansa were all carrying passengers in a variety of airplanes.

Great Britain's first airliners were modified versions of the wooden-fuselage de Havilland, Vickers, and Handley-Page bombers it had developed for the war. France too repurposed World War I technology to commercial use. In contrast, KLM Royal Dutch Airlines operated somewhat more modern Fokkers with steel-tube fuselages courtesy of the D.VII.

But the most advanced airliner after the Great War was Germany's low-wing, all-metal Junkers F 13, which featured plush seating for four in a fully enclosed passenger cabin. Building on this metal-airplane technology, Junkers introduced the Ju 52/3m, a seventeen-passenger trimotor airliner at the start of the 1930s. The Ju 52 saw wide use as an airliner by a dozen countries. As a military transport, it became the backbone of Hitler's Luftwaffe.

In the United States, meantime, the Ford Motor Company in 1925 bought up Stout, a U.S. manufacturer of single-engine transport planes built to the Junkers formula. From this acquisition sprang the Ford Tri-Motor, an eleven-passenger airliner that entered U.S. service in the late 1920s. Except for its high wing, the Ford Tri-Motor was a direct expression of the same German thinking as the Ju 52. Both airliners had a full internal metal skeleton clad in corrugated metal

With its streamlined fuselage, the Deperdussin racer achieved a speed of 125 mph (200 km/h) in 1913.

skin. They were thus enormously robust and could lift heavy loads, but they were also slow, limited in range, and expensive to operate.

The problem was that all-metal airplanes were simply too heavy when built the traditional way. Substituting steel and aluminum for wood and fabric was not the road to success. But help was on the way in the form of new ideas that had already flown. The result would be a new construction paradigm that serves aviation to this day.

A decade after the airplane was invented, the fastest and most advanced flying machine in the world also happened to be the prettiest. This was the Deperdussin racer that handily won the 1913 Gordon Bennett Cup at a blistering average speed of 125 mph (200 km/h). The single-seat Deperdussin was a high-wing monoplane with a streamlined fuselage unlike anything the world had ever seen. Built by the Société Provisoire des Aéroplanes Deperdussin (SPAD), this racing plane introduced a new way to fabricate the fuselage.

SPAD was founded in 1911 by Armand Deperdussin, a Belgian-born entrepreneur working in France. A former cabaret singer and chocolate salesman, Deperdussin made his fortune in silk trading before falling under flight's spell. Lacking a technical background, he wisely hired as his chief engineer Louis Béchereau, who handled the overall design work.

The Deperdussin racer's innovative fuselage was not Béchereau's idea, however, but rather the invention of a Swiss engineer who had previously been the Antoinette Company's shop foreman. Eugène Ruchonnet's breakthrough idea was to make the fuselage skin itself carry the loads of flight, eliminating the need for any internal support.

Ruchonnet achieved this by steaming thin strips of tulipwood, a wood favored by cabinetmakers for its pliability, and laying these softened strips into molds of each half of the plane's tapering body. Several layers of tulipwood were built up in these molds, one glued atop another with the grain running a different way for added strength. When joined together and covered with varnished fabric, these halves formed a lightweight fuselage so strong that no internal skeleton was required. Ruchonnet called this invention *monocoque* (single-shell) construction.

Where did Ruchonnet get his inspiration for so different an approach? It may have been from the small boats built in this manner or perhaps from observing the natural world. It could even be that he chanced to see a dragonfly zip past, its rigid body looking for all the world like a miniature airplane fuselage.

If small boats were in fact the inspiration, where did their creators get the idea for a rigid hull needing no internal support framework? The answer may have been the beetle. Unlike birds with their internal skeletons, beetles and many other insects rely on a hard carapace or exoskeleton—a load-bearing skin—to support their bodies and deal with the physical stresses they encounter.

Although most beetles can fly, when not in use their wings are hidden beneath hard covers that blend into the carapace. Thus it's the dragonfly with its long body and prominent wings that serves as nature's poster child for monocoque construction.

In 1913, at the height of his success, Armand Deperdussin was charged with fraud, found guilty, and sent to jail. Although he was soon released, it was a shattering blow from which he never recovered. He later committed suicide.

Fortunately for France, Louis Blériot stepped forward that year and bought up the company. He renamed it Société pour l'Aviation et ses Dérivés, thus preserving the famous SPAD acronym, and turned its running over to Béchereau, who went on to create France's most successful and widely produced fighters of World War I.

Whether in a boat, airplane, or living creature, monocoque construction works better on a small scale than on a large one. The reason is that weight increases exponentially with size. Thus, a small boat can be built as a full monocoque structure, whereas a large one requires an internal framework.

In airplanes, the point is quickly reached where a monocoque fuselage, to be sufficiently strong, must weigh more than a conventional fuselage. With traditional materials in World War I, the Deperdussin racer was about as big as one could go. The solution was to judiciously add a bit of internal bracing—a *partial* internal skeleton—to reinforce this load-bearing skin in critical places, thus allowing it to be thinner and lighter overall. This internal reinforce-

This photo of the Deperdussin's monocoque wooden fuselage marks one of the most important technological emergences in aviation.

The Albatros D.Va was one of several German World War I fighter planes with a molded-plywood fuselage.

ment was far lighter than the full skeleton required by conventional manufacture.

This modification of Ruchonnet's idea turned out to be the magic formula. Called *semi-monocoque* construction, it would be one of aviation's most spectacular successes once people figured out how to apply it to all-metal airplane manufacture. In the meantime, Roland, Pfalz, and Albatros used it to create wooden-fuselage fighter planes for Germany's war effort.

The Albatros D.Va is perhaps the best-known of these World War I fighters. Made of steam-softened plywood formed around molds, its streamlined semi-monocoque fuselage was robust and aerodynamic. However, it was also a bit on the heavy side, placing the D.Va at a disadvantage against Great Britain's Sopwith Camel and Royal Aircraft Establishment SE-5a with their conventional wood-and-wire-truss fuselages.

Between the world wars, this wooden semi-monocoque construction found memorable expression in the record-setting Lockheed Vega monoplanes of the late 1920s and early 1930s. Wiley Post circumnavigated the globe twice in his Vega, first in 1931 and again alone in 1933. Amelia Earhart used hers to solo across the Atlantic five years to the day after Charles Lindbergh.

The Lockheed Vega set numerous speed and distance records at the hands of Amelia Earhart, Wiley Post, and other noted aviators. Introduced in 1927, the Vega combined cantilevered wings with a semi-monocoque fuselage of molded plywood.

The zenith for wood came in World War II. For more than two years, the de Havilland Mosquito fighter-bomber—Great Britain's "Wooden Wonder"—was the fastest thing in the skies over Nazi-occupied Europe. So quick were bomber versions of the Mosquito that they dispensed with guns and relied on speed alone for protection.

De Havilland built the legendary Mossie out of wood because it anticipated a wartime shortage of aluminum that never materialized. In the United States, this same false expectation gave rise to the Hughes H-4 Hercules, the giant one-of-a-kind flying boat built by Howard Hughes as a World War II cargo plane. Not finished until two years after the war had ended, the *Spruce Goose*, as this huge machine is known, lifted off the water just once in a brief straight-ahead hop in 1947 with Hughes himself at the controls.

Despite its wartime success, wooden semi-monocoque construction was a technological dead end long before World War II. It petered out right after the war with the de Havilland Vampire and Venom jet fighters—yes, wooden jets.

Back in the 1920s, consensus had already emerged that metal was aviation's future. In addition to its greater strength and predictability,

Designed for strafing trenches, the armored Junkers J 4 of 1917 was history's first production airplane made entirely of metal.

metal could be formed and fastened in more ways, and it was more resistant to moisture and temperature extremes. The only trouble, in fact, was that the world kept trying to build metal airplanes the wrong way. It took time for people to understand that semi-monocoque construction (also called *stressed-skin* construction) was the correct way to exploit the potential of steel and aluminum.

Also in the 1920s, Germany overtook France as the world leader in flight technologies. It was in Germany that much of this all-metal construction puzzle was figured out. However, it would not all come together until the start of the next decade in the United States, whose own technological ascendancy was just beginning.

Skimming the trenches in 1917, the low-flying German warplane sprayed out a hail of machine-gun fire. Soldiers quick enough to shoot back saw the odd craft continue unfazed on its ugly mission. They had encountered the Junkers J 4, the world's first operational airplane made entirely of metal.

Spindly, unwieldy, and so heavy it could scarcely climb, the J 4 nevertheless found use in the ground-attack role because its steel and duralumin structure, locally enhanced with armor plating beneath to protect its engine and crew, rendered it less vulnerable to ground fire than conventional wood-and-fabric airplanes.

Professor Hugo Junkers, a towering figure in the development of flight.

The all-metal J 4 was the brainchild of Hugo Junkers, an industrialist who before the war had designed and manufactured water heaters. More recently, Junkers had served as a college professor at Aachen's famous technical university. He thus combined theoretical engineering knowledge with hands-on expertise in the fabrication of metal structures.

From this unusual background sprang the idea of airplanes built entirely of metal. Junkers, a socialist and pacifist, brought this proposal forward with little thought to how it might be employed. To his distress, he found himself forced under the excuse of wartime urgency to pursue a deadly application of this vision.

The first result had been the Junkers J 1 *Blechesel* (sheet-metal donkey) of 1915, a rugged steel monoplane so heavy that it accelerated slowly and barely climbed at all. Junkers progressed to a lighter design built partly out of duralumin, an early aluminum alloy highly subject to corrosion. Because warplanes weren't expected to last, this was not seen as a disadvantage.

A passionate advocate for the monoplane, Junkers designed thick wings that did not require external bracing struts or wires. This was too radical for the military authorities, who instead ordered him to prepare a biplane for production. To help with the production side of things, they brought in Anthony Fokker, the young Dutchman whose company was Germany's premier supplier of fighter aircraft during the war.

The result of this temporary collaboration was the trench-strafing J 4, an angular two-seater also confusingly known as the Junkers J 1 (the former was the company's name for the airplane, the latter its military designation). The J 4 was actually a *sesquiplane*, which is a biplane with one wing significantly smaller than the other (*sesqui-* means "one and a half," and *plane* refers to a lifting surface or wing).

Following Fokker's departure, Junkers went back to a single wing for the all-metal D 1, a low-wing fighter plane with corrugated skin. He also delivered the CL 1, a two-seat version with a second cockpit for a rear-facing gunner. From these two 1918 machines sprang the postwar F 13 of 1920—an astonishing airliner described later in this book—and subsequent Junkers transport planes of the interwar era.

Junkers was a pioneer with good ideas for building airplanes out of metal. For example, he backed smooth sheet metal with corrugated sheets for high strength and resistance to buckling. He also understood that the skins of metal airplanes could shoulder some loads, which is why the Junkers Ju 52 and Ford Tri-Motor have corrugated aluminum cladding. For aerodynamic reasons, these corrugations are aligned parallel to the airflow.

Ultimately, though, Junkers missed the boat when it came to building out of metal. Locked into the reigning paradigm, he failed to question the assumption that metal airplanes must carry their structural loads the way wooden ones had, or for that matter the way vertebrate animals do. As a result, his designs retained an internal skeleton.

Making metal airplanes this way was a bit like dressing a medieval knight in a suit of armor and ordering him to go do a regular day's work despite the added weight. A compatriot of Junkers' would be the first to realize this.

Famous for its dirigibles during World War I, the Zeppelin company also built large bombing airplanes at its Staaken plant in the Berlin suburbs. Dr. Adolf Rohrbach, a remarkable young engineer, was the designer of those Staaken *Riesenflugzeuge*, or giant airplanes.

After the war, Rohrbach set about designing an all-metal airliner with seating for up to eighteen passengers. This machine would advance the state of the art by combining what he and his colleagues had learned about large-airplane design with knowledge gleaned from Zeppelin's pioneering use of aluminum in dirigibles. The result was the Zeppelin-Staaken E.4250, history's first all-metal airplane to employ stressed-skin construction. Completed in the fall of 1920, the E.4250 (also known as the E.4/20) was astonishingly modern-looking, with an aerodynamically clean fuselage and four engines inset in a minimally braced high wing.

In addition to semi-monocoque construction, this advanced airliner featured counterrotating propellers to eliminate asymmetries of thrust (the propellers on one wing turned in the opposite direction from those on the other). In fact, the design's only retrograde features were a fixed landing gear and a dreadful cockpit location. Instead of sitting at the front, as one would expect, the E.4250's pilot sat high atop its whale-like back and between the wings, where he couldn't see much of anything.

This was an enormous machine with a maximum gross weight of almost 19,000 pounds (8,500 kilograms) and a wingspan of about 102 feet (31 meters). Although it flew well in tests, the Zeppelin firm saw its hopes for production dashed by the Inter-Allied Commission of Control, which—noting the craft's military pedigree and potential—ordered it destroyed under the terms of the Treaty of Versailles. Sadly, despite the company's impassioned pleas to be allowed to sell or donate the E.4250, it was scrapped in November 1922. Had this singular airplane instead entered commercial service and spawned successors, all-metal semi-monocoque construction might have blossomed earlier.

The Zeppelin-Staaken E.4250 of 1920 might have hastened the adoption of all-metal, stressed-skin construction had it not been ordered destroyed under the terms of the Versailles Treaty.

As for Adolf Rohrbach, he relocated to Denmark to avoid Versailles Treaty limitations and opened an airplane plant that built land planes and flying-boat airliners for Lufthansa. He relocated to the United States at the end of the 1920s and returned to Germany the next decade, where he died on the eve of World War II at age fifty.

Who did first put it all together? The honor of designing the world's first production all-metal, semi-monocoque airplane goes to John K. "Jack" Northrop in the United States, with his Alpha. Northrop, a self-taught aeronautical engineer, is one of the most influential airplane designers of all time. Working at various times for companies such as Lockheed and Douglas, Northrop always wanted his own company and had one at different stages of his career. Best known for the airplanes that bear his name, he also helped shape other people's designs, ranging from Lindbergh's Ryan NYP *Spirit of St. Louis* to the famous Douglas DC-3 airliner.

An avid follower of aviation trends, Northrop came up with the idea for an aerodynamically clean and very fast high-wing cabin monoplane in the mid-1920s. The result was the Lockheed Vega, which embodied Northrop's ideas for semi-monocoque construction combining pressure-molded wooden halves with an internal framework. Sleek and capable, the Vega was an instant success that only

The Northrop Alpha of 1930 was the first production airplane to employ all-metal, semi-monocoque construction.

improved as more powerful engines and aerodynamic cowlings became available. But Northrop was now thinking in terms of metal, not wood.

In 1928, Northrop left Lockheed to produce the first airplane to bear his name. The Northrop Alpha flew in early 1930. Sleek and shiny from its fully cowled engine to the tip of its tail, this breakthrough mail plane combined a fully cantilevered wing with an enclosed cabin for airmail, freight, or up to six passengers. Coming at the start of a new decade, the prototype Alpha was a transitional monoplane with holdover biplane design shortcomings. First, it had an open cockpit behind its enclosed cabin that placed the pilot far aft, where visibility was poor. Second, it lacked a retractable landing gear, although production Alphas were fitted with aerodynamic gear fairings to reduce drag.

Just what is stressed-skin construction? Imagine you're holding a thin sheet of aluminum about a meter square. This piece of sheet metal is very light. Asked to try your strength on it, you confirm it's too strong to be pulled apart. You also find you can't distort its square shape into a parallelogram. In engineering terms, this material has just demonstrated excellent resistance to *tension* and *shear* loads.

Now, donning heavy gloves to protect your hands, you push inward on its sharp edges. As you expected, the thin sheet metal buckles easily. You find that you can fold, dimple, or otherwise permanently distort it. In fact, although it is a bit thicker and stiffer, this sample of airplane skin reminds you of the empty aluminum soft drink cans you regularly recycle.

You have just found aluminum sheeting's Achilles' heel: poor resistance to *compression* loads. Thick aluminum plate could resist these loads, but that would make for an airplane too heavy to fly. Here was a key structural challenge for those seeking to build out of metal.

The genius of semi-monocoque construction is that it calls on thin aluminum to carry the loads it *can* withstand (tension and shear) and spares it the one it can't (compression). It does this by clever use of an internal framework far less substantial than a full load-bearing skeleton. By itself, what's beneath the skin of one of today's all-metal jetliners could not begin to support the craft's weight on the ground, let alone the loads imposed by flight.

As worked out by Rohrbach, Northrop, and others, this internal framework comprises *circumferential members* (ring formers and occasional bulkheads) with fore-and-aft or *longitudinal members* (longerons and stringers). Together these elements create sturdy compression-resistant bays over which aluminum skin is pulled taut and riveted in place. The result is a light yet strong metal structure in which the skin and what's behind it share the job of carrying the loads and stresses of flight. Channel sections, flanges, built-up assemblies, and other bits of engineering magic further enhance strength while keeping overall weight to a minimum. Advancements in metallurgy, fabrication techniques, stress modeling, and structural design have continued to improve this breakthrough construction method over the decades.

About the same time that Jack Northrop's company flew the Alpha, Boeing in Seattle flew its first all-metal, semi-monocoque airplane. The Boeing 200 Monomail was a mail plane of similar over-

Carrying ten passengers at 180 mph (290 km/h), the Boeing 247 of 1933 was the world's first modern airliner.

all configuration except that it featured a retractable landing gear. Although the Monomail wasn't produced, it immediately gave rise to other Boeing stressed-skin airplanes that were.

The culmination of this cutting-edge activity in Seattle was the Boeing 247, a ten-passenger airliner that took wing in 1933. In terms of configuration and construction, the Model 247 marks an enormously significant historical emergence: three decades after the Wright brothers invented the airplane, here was the world's first modern passenger airliner. In contrast to other transports of the day, it alone possessed the sleekness of semi-monocoque construction.

The Model 247 also boasted retractable wheels, the latest radio gear, gyro instruments for blind flying, and other cutting-edge technology. Unfortunately, though, the airplane itself carried too few passengers to find solid commercial success. Moreover, passengers didn't like the wing spars invading its passenger cabin, which created obstacles they had to climb over to reach their seats.

Douglas Aircraft saw a chance to do better. The next year, they began deliveries of the DC-2, a fourteen-passenger airliner with all-around better performance than the 247 as well as a larger cabin that

The Douglas DC-3 forever changed the world when it entered service in 1936.

sat above the wing spars so its aisle was uninterrupted. But it was this California company's next product that changed the world.

Placed into service in 1936, the Douglas DC-3 would be the most dominant airliner ever and one of the most significant airplanes in history. A bit bigger all around than the DC-2, it had a wider fuselage that allowed for ten railroad-style Pullman berths on night flights. In a day-plane configuration, this cabin accommodated twenty-one passengers seated three abreast, two on the left side of the aisle and one to the right.

Here at last was a magic combination of payload, performance, and ruggedness that let airlines earn solid profits even without subsidies. The air-transport industry grew up with the DC-3, which by 1939 accounted for an astonishing 90 percent of the U.S. commercial fleet and also flew with two dozen carriers around the world.

On the eve of World War II, the Douglas DC-3 showed what semi-monocoque construction could do. Douglas delivered about 420 DC-3 airliners before the war halted production. It then rolled out 10,000 military variants for use by U.S. and other Allied forces. Yet more were manufactured under license by both the Soviet Union and Japan. All these decades later, hundreds of these rugged transports are still at work hauling cargo in many countries around the world.

World War II provided astonishing endorsements of the strength of stressed-skin construction. No airplane bore more dramatic witness to this than the Boeing B-17 Flying Fortress, which ranks as one of the most rugged airplanes of all time. Wartime photographs depict B-17s returning with the entire nose section blown off, gaping holes in the wings and fuselages, most of the tail gone, and so on. In one famous instance, a Flying Fortress returned sliced nearly in half by the wing of a colliding German fighter.

Of course, success breeds hubris, and with hubris there is often a fall.

On May 2, 1952, the de Havilland Comet I—history's first jetliner—entered commercial service between London and South Africa. Sleek and futuristic, it looked to be a winner until two fatal crashes showed that people had more to learn about all-metal airplane design.

The de Havilland DH.106 Comet entered service in 1952 as the world's first jetliner. The prototype, shown here, flew in 1949.

On January 10, 1954, a Comet I disintegrated in flight off the coast of Italy. Finding nothing wrong with the design of the airplane, investigators tentatively attributed the tragic event to a bomb. But then on April 8 another Comet mysteriously came apart over the Mediterranean, and a more sinister picture emerged. The Comet fleet was grounded and an investigation of unprecedented scope was launched. In retrospect, people now suspected that a third crash in India, formerly attributed to a violent tropical storm, might also be a case of spontaneous structural failure in flight.

An unseen killer was stalking the Comet, but what was it? Engineers and accident investigators worked around the clock exploring every possibility. Finally, pressure testing of a Comet fuselage in a water tank together with analysis of recovered parts identified the culprit: metal fatigue.

The Comet flew twice as high as most other pressurized airliners of the era, so its pressurized cabin sustained a higher pressure differential between inside and outside. This exposed its metal body to stress every time the cabin was pressurized. The resultant flexing weakened the structure until it failed in a catastrophic rupture at altitude.

Suspicion initially focused on an escape hatch in the roof of the flight deck, which appeared to have failed. Wreckage subsequently brought up by a fishing boat suggested it might instead have been a passenger window. But exactly where the failure occurred was moot; the important thing was that it was occurring at all.

With little prior experience in pressurization, de Havilland's engineers had not known to avoid designing windows and hatches with square corners. Unlike round metal window frames, or even square ones with rounded corners, square frames concentrate stresses at their corners. This alone had not doomed the jets, however, because these were robust structures. All calculations suggested they were sufficiently strong.

Neither de Havilland nor Great Britain's certification authority, which had approved the Comet for production, was thus responsible for what had happened. Instead it was a case of simply not enough being known at the time about how metal fatigued under repeated stresses such as pressurization cycles.

Armed with the hard-won knowledge of bitter experience, de Havilland set about completely redesigning the Comet. Now fitted

With the 707 of 1958, Boeing launched a rapid transition to turbine propulsion.

with oval windows and stretched to carry ninety passengers, it returned to service in October 1958. By then, though, it was too late; less than two weeks later, the Boeing 707 also entered service. Far more capable, the 707 was the plane that ushered in the commercial jet age.

Engineering is a discipline that learns from failure as well as success. The Comet's audacious leap and subsequent fall at least helped lift the industry to a higher plateau of safety. So too did the 707, whose success rested on its maker's world-leading philosophy of safe airplane design.

Among other things, that uncompromising Boeing philosophy prohibited single-failure modes, which in plain English means that no failure of any individual system or structural component can ever be permitted to endanger the airplane or its occupants. Consequently, designs are required to be robust and employ redundancy where needed in the form of backup systems and alternative structural load paths.

A spectrum of simple yet powerful design ideas such as this one has made air travel vastly safer over the decades. Another example is a blanket aviation industry design prohibition against uninspectable limited-life components. The idea here is that if a key structural part is located where an airline can't get to it to check it during maintenance, then the part must be made to last the life span of the airplane. Degradation that could affect airworthiness is not permitted.

Despite a host of improvements over the decades, airliners today are built basically the same way that the Boeing 247 and Douglas DC-3 were back in the 1930s. All-metal semi-monocoque construction has served the world well.

The newest jetliner at the time of this writing is the Boeing 787 Dreamliner, which features an innovative plastic body that is much stronger and lighter than metal. Despite having a composite primary structure, however, the Dreamliner remains a semi-monocoque airplane and is thus the dragonfly's conceptual progeny.

5 WINGS, PART I

FROM BOX KITES TO BRIDGES

> In the early days, the chief engineer was very often also the chief test pilot. This tended to result in the elimination of poor engineering.
>
> —IGOR SIKORSKY (1889–1972)[1]

Engines rattling, the Sikorsky Ilya Muromets trundled down the field at St. Petersburg, Russia, and climbed slowly into the air. Dipping its wings to one of Europe's most beautiful cities, the enormous biplane pointed its nose southward.

It was June 21, 1914. Less than five years after Louis Blériot staggered across the English Channel in a frail 30-hp single-seater, here was a flying giant with an enclosed cockpit with dual controls, a plushly furnished passenger cabin, four engines of 148 hp each, and a wingspan of 102 feet (31 meters).

At the helm of this astonishing giant was its twenty-five-year-old designer, Igor Ivanovich Sikorsky, a figure who ranks just behind the Wright brothers themselves in the annals of early flight. Chief engineer of the Russo-Baltic Carriage Factory, a company specializing in railcar manufacture, this humble genius was also its aviation shop foreman and test pilot.

Just twenty-three years old, Igor Sikorsky sits at the controls of his Grand, the world's first multiengine airplane, in 1913.

A year earlier, Sikorsky had built and flown the Grand, history's first multiengine airplane. When that behemoth was destroyed in a freak accident (the engine fell off a landing Morane airplane and plummeted through the parked Grand), this brilliant Ukrainian had built an even bigger and better machine.

Named for a mythical Russian folk hero of the tenth century, the Ilya Muromets made its first flight the previous December, a week shy of the tenth anniversary of the Wright brothers' crowning success. In February, it had carried sixteen people aloft at once. Now on the first day of summer it was off making another record attempt: a cross-country flight to Kiev, Ukraine, with one stop each way for fuel and rest.

Climbing out after the refueling stop at Orša, Sikorsky corrected for constant thermals that jounced his machine like a bad road. Updrafts filled the cabin with the sweetness of sun-warmed meadows and

A decade after the Wright brothers' success at Kitty Hawk, the four-engine Sikorsky Ilya Muromets took wing in Russia.

the heady redolence of flowering hedgerows. Trees gave way to grazing cattle and astonished field hands as Sikorsky followed the land's contours. Squinting into the sun, he let the Dnieper River lead him south to a hero's welcome at Kiev, his hometown.

The reverse course was flown the following day. That evening, having surmounted mechanical troubles and bad weather, the crew of the Ilya Muromets was home again in the city of Peter the Great. They had covered 1,400 miles (2,200 kilometers) in four stages, putting the rest of the aviation world to shame.

Here in imperial Russia's capital in 1914 was what George Cayley had imagined and William Henson had tried to build: a practical airliner offering genuine utility. It would have written a different beginning to commercial air travel had world events not intervened.

A week after Sikorsky's flight, a hotheaded young Serbian nationalist assassinated Austria's Archduke Franz Ferdinand, presumptive heir to the Austro-Hungarian throne, and his wife, Sophie, in the streets of Sarajevo. That act of violence triggered a series of military escalations across the European continent that by August erupted into open warfare. World War I had begun.

Instead of building airliners, the Russo-Baltic Carriage Factory would roll out more than seventy Sikorsky Ilya Muromets bombers. They found limited use because the Great War needed smaller, short-range tactical bombers to support military operations right at the front, not big strategic ones capable of flying far beyond enemy lines.

Preoccupied with the longer and more active Western Front, the Allied powers paid scant attention to the war in the east. Not so the Germans who met Sikorsky's amazing giants in the air or shuddered under their bombs on the ground. The German *Riesenflugzeug* (giant airplane) program of World War I was the direct response to those lumbering marvels.

How did Igor Sikorsky come up with the amazing Ilya Muromets? The answer lies in part in this airplane's extremely long wings. Sikorsky knew that slender wings of great span lift more efficiently than broader, shorter wings of the same total area.

In fact, there were many secrets just waiting in the wings.

Birds have thin wings, so that was where flight's pioneers began. But how did one go about constructing a scaled-up bird wing that could support a flying machine? It was a huge challenge, one exacerbated by scale effects and the constant need to keep weight to a minimum.

The construction of ships, buildings, and bridges provided adaptable engineering knowledge along with skills, tools, and techniques in woodworking, metal fabrication, and other disciplines. In the patent drawings for his 1843 Aerial Steam Carriage, trained mechanical engineer William Henson called on all of them as he focused his creative energies skyward.

The wings that Henson designed for the Aerial Steam Carriage comprised three beams extending from each side of the fuselage. These lateral *spars* were the wings' primary load-bearing members. He crossed them with fore-and-aft members called *ribs* that created a structural lattice and gave the wings their cambered airfoil shape.

Thus bound and internally reinforced with cross-bracings, these spars and ribs together formed lightweight wing panels. Henson's drawings called for them to be encased in cloth treated to prevent air from penetrating its weave. Henson specified oiled silk or, if that was too expensive, canvas.

Wing panels constructed this way are light and hold together well, but in one key regard they are woefully deficient: they cannot resist twisting or *torsional stresses* down their length. Rotated by gusts in flight, the wingtips would angle up or down, destabilizing the airplane and flexing the wing until it failed and snapped off. Long before that could happen, though, the wing's erratic aerodynamic performance probably would have flown the plane into the ground or a tree.

To see why this is the case, imagine that you pass through the door of an unfinished house and find yourself standing on a long, narrow section of wooden flooring. This floor is structurally similar to Henson's wing panel because its fore-and-aft joists serve the same function as the wing spars while its transverse floorboards act structurally like wing ribs.

Wooden floors built this way are safe to walk on because their joists are supported at both ends. Imagine for a moment, though, that this particular section of flooring—a narrow swath encompassing two joists—is supported at only one end. The joists come out of the wall below you and extend forward, and the floorboards run from left to right.

As you walk out from the wall, you find that this freestanding span supports your weight all the way out to the tip. As your weight reaches the unsupported end, you find it scarcely bends beneath you like a diving board. The reason is that bound joists and floorboards form a beam structure that resists fore-and-aft or *spanwise* flexing.

Unfortunately, though, you also discover that the farther you go from the wall, the more this freestanding section of floor twists and tilts to one side or another under your body weight. Only by holding to the center and balancing carefully can you keep this lateral flexing from sliding you off to the side. The reason is the poor resistance to chordwise flexing of thin, cantilevered panels, be they wings or flooring.

Here in a nutshell was the thorny problem confronting Henson and the other early wing designers: while bird-like wings could be made that wouldn't flex or snap off, they could not be made resistant to twisting. Henson's design solution was to use *external bracing wires* strung along vertical posts mounted on the aircraft's wings and fuselage. These bracing wires ran all the way out to the tips of the Steam Carriage's wings to keep them rigid. Still more wires reinforced the tail.

Henson's engineering was a little squirrelly, but basically he was on the right track. Of course, none of this—not beams, spars, or trusses—was new, as one glance at a sailing ship will confirm. Henson merely appropriated existing pieces of maritime technology in his effort to sail the heavens.

Aviation would put this idea of external bracing wires to good use, as the Antoinette, Blériot XI, and other pre–World War I monoplanes show. But the world's first successful airplanes would be biplanes, not monoplanes, and the reason takes us Down Under.

Lawrence Hargrave first saw Australia as a boy of fifteen. The second son of an immigrant British family, he was an adventurous youth who spurned conventional studies. Instead he pursued a marine engineering apprenticeship that let him join maritime expeditions of discovery.

His new world was a panoply of tropical heat, groaning timbers, and billowing sails. Porpoises breached the pristine seas in quicksilver arcs. Violent squalls descended with little warning. Waves crashed loudly over coral reefs, their foam visible even by moonlight.

All of it fascinated Hargrave, whose interest in the natural world recalls that of Charles Darwin aboard the HMS *Beagle* earlier that century. In addition to circumnavigating Australia, Hargrave visited New Guinea several times, exploring its coastal waters and headlands. It was a hard and dangerous life.

On one voyage, a storm set the ship adrift by tearing off its rudder. To the relief of all, Hargrave improvised a tiller out of a capstan. Then a worse storm ripped the luckless ship asunder, claiming many lives. Hargrave survived by clambering desperately up a mast as the ship slipped beneath the waves. A lifeboat risking all in the heaving seas plucked him to safety.

In 1876, this young man joined an expedition of discovery aboard a small steamship that made its way 400 miles up a New Guinea river. Hemmed ever closer by dense rain forest, the crew survived attempted canoe ambushes and hails of arrows. Hargrave and his companions saw indigenous peoples who lived in swaying tree dwellings. They also saw inhabited riverside huts festooned with shrunken heads and painted skulls.

Throughout those adventurous six years in his twenties, it was the birds that fascinated Hargrave the most. He loved their calls, their brilliant plumage, the way they rode the air. Because if there was one thing he believed, it was that human beings would soon teach themselves to fly.

Returning to Sydney, Lawrence Hargrave settled down, married, and started a family. His fascination with the natural world led him to the Royal Society of New South Wales and duties as an assistant astronomer with the Sydney Observatory. But his thoughts were never far from flight, and when his father died some years later, leaving him with independent means, he left that post to devote himself full-time to aeronautical studies.

Hargrave began by observing birds and soon progressed to experimenting with model gliders and kites. To this end, he moved his growing family to Stanwell Park, a coastal community south of Sydney that offered steep slopes, a sandy beach, and steady winds. Here Hargrave independently rediscovered what Cayley had learned eight decades before: curved surfaces, such as a bird's wing, produce more aerodynamic lift than flat ones.

Lawrence Hargrave.

In 1893, this continuing investigation culminated in Hargrave's great gift to

In November 1894, Hargrave ascended into the air beneath a train of box kites.

aviation: the box kite. This new type of kite lifted strongly, was very stable, and was structurally robust. In it Hargrave rightly perceived the basis for a man-carrying flying machine.

Constructed of cross-braced parallel struts, the box kite was open at its ends and wrapped with taut fabric around its cells. Using an anemometer to determine wind speed, an inclinometer to measure the kite string's angle, and a spring balance to determine its lift, Hargrave carefully noted his invention's performance and characteristics through successively refined versions.

On November 12, 1894, he demonstrated his faith in box kites by lifting himself 16 feet (5 meters) into the air supported by four of them strung in train. In a scientific paper published the following year, he cited this success as proof that "an extremely simple apparatus can be made, carried about, and flown by one man; and that a safe means of making an ascent with a flying machine, of trying the same without any risk of accident, and descending, is now at the service of any experimenter who wishes to use it."[2]

Hargrave did not patent his invention because he wished the world to have free use of it for the betterment of all. In 1899, he traveled to London, where he proudly demonstrated it. Europe's flight experimenters immediately embraced the box kite for its combination of high lift, inherent stability, and structural efficiency. To see just how influential this Australian's thinking was in Europe, one has but to glance at Alberto Santos-Dumont's 14-*bis* of 1906; it resembles nothing so much as a collision of Hargrave box kites hitting at different angles.

But it was in America that Lawrence Hargrave's profound idea would first flower.

Human thought flows freely across natural barriers and national boundaries. Hargrave's correspondence with fellow flight enthusiasts in the British Isles, Europe, and North America began circulating his ideas, which would contribute to flight's invention.

Two Hargrave correspondents in North America were keen to fly. One would not know what to do with the Australian's invention; the other would actually improve on it. The first was Alexander Graham Bell, the telephone's famous inventor.

Bell had emigrated from Scotland to Canada in 1870, at age twenty-three. Crossing into the United States the next year, he eventually became a U.S. citizen but continued to summer in Canada's Nova Scotia. There Bell took up aerial experiments with large kites made of interlocking tetrahedrons. His interesting ideas on construction created kites that lifted heavy loads, but they could not live up to his ill-defined hopes for a novel flying machine that was stable, could fly slowly, and was capable of human transport.

In 1907, Bell made his one contribution to aviation, albeit an indirect one, by chartering a small group that brought younger blood to the challenge of flight. Called the Aerial Experiment Association, this team's star was Glenn Curtiss, who would take the grand prize at Reims in 1909 and become America's most successful early airplane manufacturer.

Fortunately for the Wright brothers, Lawrence Hargrave's other American correspondent *did* know how to employ the box kite. Retired U.S. railroad engineer Octave Chanute was a flight devotee who for years had applied his structural design and stress analysis expertise to aerial speculations. These he published as a bound set in 1894 under the title *Progress in Flying Machines.*[3]

A prolific correspondent and internationally acclaimed lecturer, Chanute had amassed considerable wealth and an international reputation building bridges and other key pieces of the young United States' developing infrastructure. He loved to keep tabs on flight-related research and experimentation around the world, and in retirement served as a self-appointed clearinghouse for information. Much of what he learned found its way into his influential book.

Generous with encouragement, Chanute cross-pollinated others' efforts with infusions of new ideas, his own and those of others. Often he visited these isolated pockets of experimentation to observe their efforts firsthand. To those he considered worthwhile, he even occasionally provided financial assistance.

Chanute was particularly impressed with Hargrave, whom he came to know through letters. When Hargrave described the box kite to him, Chanute instantly appreciated this invention's potential. "If there be one man, more than another, who deserves to succeed in flying through the air," he proclaimed some years before meeting the Wrights, "that man is Mr. Lawrence Hargrave of Sydney, New South Wales."[4]

That support meant a great deal to Hargrave because not everyone in Australia shared his belief in aviation. "The people of Sydney who can speak of my work without a smile are very scarce," he admitted to Chanute. "I know that success is dead sure to come, and therefore do not waste time and words in trying to convince unbelievers."[5]

Now it was the eventful summer of 1896. Otto Lilienthal was gliding in Germany, Samuel Langley's steam-powered model had flown a kilometer in May, and in general a surging sense of optimism prevailed that the challenges of manned, powered, heavier-than-air flight might actually be solved.

Chanute shared Lilienthal's view that unraveling aviation's secrets required actually attempting to fly. With this in mind, he convened a group of younger enthusiasts and hosted them at the southern tip of Lake Michigan in Indiana, some 30 miles southeast of his Chicago home. Living in tents in that pristine area (soon to become the steel-mill city of Gary), this group tested a variety of gliders.

In his mid-sixties, Chanute left the flying to others and looked on. He was keenly disappointed when his Katydid glider, with its six tiers of pivoting wings, proved an outright failure. With it went the civil engineer's hopes of automatic stabilization in flight. Other more conventional craft, including a Lilienthal glider copy flown by young engineer Augustus Herring, flew passably well but contributed no new knowledge.

In August word arrived of Lilienthal's death. This sad news from overseas did not stop the Chanute team's forays to the Indiana dunes, which continued into the fall. This gentlemanly experimentation culminated in a significant glider developed collaboratively by Chanute and Herring.

The Chanute-Herring glider of 1896 combined Lawrence Hargrave's box kite idea with Octave Chanute's engineering expertise to create biplane wings.

By all accounts, the Chanute-Herring machine surpassed Lilienthal's best achievements. A trim little design, it was essentially Hargrave's box kite reimagined with a cruciform tail and wings modified for its new role as a man-carrying glider. But Chanute's embellishments were what make this human artifact so significant historically.

Chanute's engineering expertise imbued Hargrave's breakthrough with a combination of struts and wire bracing that trussed the two wings into a single rigid beam structure. Neither wing alone could have withstood the twisting and flexing forces of flight, but bound together, each lent the other support to create a structure as sturdy as a packing crate.

Here was aviation's equivalent of the Pratt truss, a method of distributing and supporting heavy loads popular with nineteenth-century civil engineers. The externally braced biplane—history's first heavier-than-air flight structure—had come into being.

The wings are the most import part of any airplane. They largely determine its performance capabilities and flight characteristics. This is true regardless of the airplane's configuration.

Biplanes and monoplanes are alike in this regard but differ from a load-bearing standpoint. In biplanes, the wings are the primary structure; all else relies on the rigid beam truss created by these braced lifting surfaces. In contrast, the fuselage is the primary structure in a monoplane because it supports the wings and all else.

In creating the biplane, Chanute had taken Hargrave's brilliant idea of mutually reinforcing wings and added a masterly understanding of structural load paths. What he had learned over decades of designing bridges and other structures, he now translated aloft. The result was a strength-to-weight ratio that placed manned, powered, heavier-than-air flight well within grasp.

The first beneficiaries of Chanute's repurposing of the box kite were none other than Wilbur and Orville Wright. In the Chanute-Herring glider, the mechanically minded brothers recognized a design solution combining structural rigidity, light weight, and ample wing area.

Aside from encouragement, this was Chanute's single contribution to the success of the Wrights. The resemblance of their gliders and flyers to his seminal 1896 machine shows how faithfully the Ohio brothers took its lessons to heart.

In one regard, however, the Wrights would depart from what Hargrave and Chanute had achieved. In a stroke of genius, they deliberately weakened this rigid wing truss to allow the wings to twist for control in the air, as we shall see in a later chapter.

From the time he was tiny, Geoffrey Hargrave had helped his father with his flight experiments. When the new century arrived and airplanes emerged, father and son pored over the press accounts in joyous celebration. It was a matter of pride that the world's first heavier-than-air flying machines bore the Hargrave stamp.

Then World War I broke out. Australia and New Zealand, dominions of the British Empire, dutifully sent their young men off to fight. Geoffrey Hargrave volunteered and was among the Anzac infantrymen whose lives were senselessly squandered at Gallipoli in Great Britain's disastrous Dardanelles Campaign. The Turkish bullet that ended young Geoffrey's life in 1915 broke the elder Hargrave's heart. Australia's great aviation pioneer succumbed that same year at age sixty-five.

Even as the structural demands of wings were being figured out, people were also probing their aerodynamic qualities. This began with George Cayley, who postulated for the first time in history that curved surfaces lift more effectively than flat ones. The father of wing theory, Cayley used a whirling arm to methodically conduct the world's first airfoil tests.

In 1804, George Cayley came up with a fascinating idea. If the wings angled upward as they extended from the center, would that make an airplane self-righting from side to side? Glider tests quickly confirmed this hunch. Cayley had hit upon wing dihedral, a slight upward tilt to the wings that increases lateral stability. Dihedral also fights against maneuverability, however, so airplane designers use it sparingly, if at all.

Most airplanes, past and present, have a few degrees of dihedral to reduce pilot workload. High-wing designs generally require less dihedral than low-wing ones because of the position of the airplane's center of gravity relative to its wings.

Some airplanes have no dihedral, and some even have negative dihedral or *anhedral.* A downward tilt to the wings, anhedral makes an airplane inherently unstable but highly maneuverable. The Wrights chose to build with anhedral that was obvious when the Flyer was on the ground but less so in flight, when aerodynamic lift brought the wings almost level. This design decision made their airplanes less sensitive to gusts of wind from the side as well as more maneuverable.

The next time you're at an airport and see a jetliner, take a look at its wings. You'll notice this slight upward tilt and know why it's there.

Scientists may discover the physical universe's secrets, but engineers are the ones who change our world. In the early days of flight, these disciplines frequently overlapped, to aviation's great benefit.

Francis Herbert Wenham was a theoretician with a foot in each camp. Born in London in 1824, this mechanically gifted son of a British Army surgeon grew up building and flying kites. Pursuing university studies, he emerged with a degree in engineering and wide-ranging interests.

Wenham's face was jovial and his eyes observant. His beard and hair were often in wild disarray. Energetic, fascinated by the natural world, and keen on science, he would contribute in his lifetime to fields as diverse as photography, instrument design, microscopy, and high-pressure steam engines.

FIG. 101. FRANCIS HERBERT WENHAM.

Francis Herbert Wenham.

On a trip to Cairo in his thirties, this Englishman found himself enchanted by the birds that flocked noisily overhead and alighted to feed in the Nile wetlands. Egypt's location at the crossroads of three continents made this river's verdant course through the desert a migratory corridor for millions of birds. There were swifts, kites, bulbuls, kingfishers, swallows, gallinules, and other species Wenham couldn't identify.

The engineer in him marveled at the marked differences in how various kinds of birds flew. Even as his eyes assessed their abilities, his mind began equating them with the shapes of their wings. Back in the British Isles, Wenham undertook studies in this area. He was reportedly an excellent shot and collected specimens while on upland bird hunts.

Wenham was familiar with the work of George Cayley. Building on the latter's advocacy for cambered lifting surfaces, Wenham observed that all bird wings are thicker near the leading edge. To engineers such as Wenham, thicker also means stronger. This led him to theorize that wings derive most of their lift toward the front. Why else would nature construct them this way?

This key insight dovetailed with Wenham's observation that the birds with the longest wings were the strongest fliers. Short wings afforded birds greater maneuverability, but long ones kept them aloft longer and with less effort.

To test his insight, the enthusiastic Englishman decided to design and build a manned glider. Beyond Cayley's insights and configurational advice, however, he could find virtually no information to help him with this self-appointed task. It was only the start of the 1860s; another three decades would pass before Otto Lilienthal would solve similar challenges.

Wenham ended up constructing a glider with five sets of wings mounted one above the other like slats on a venetian blind. He felt he needed this many wings because each was so narrow from front to back. He built them that way on purpose so that proportionately more of the glider's total wing area would be close to the wings' leading edges, where he believed most of the aerodynamic lift would be created.

Unfortunately, this line of experimentation proved an outright failure. The glider wouldn't fly, and the whole frustrating experience drove home to Wenham just how little was actually known to help experimenters such as himself. Fortunately for posterity, he did something about it.

In 1866, he became a founding member of the Aeronautical Society of Great Britain, predecessor to today's Royal Aeronautical Society. Created to help advance the quest for human flight, this engineering society boasted among its membership many of the leading scientific and engineering luminaries of the day.

Called on to give the society's first annual lecture, Wenham presented a paper titled "On Aerial Locomotion and the Laws by Which Heavy Bodies Impelled Through Air Are Sustained." A masterly blend of observation and deductive reasoning, this talk addressed bird flight, assessed the state of flight research, identified and examined technical issues, advocated for kite and glider experiments, and presented speculations on glider design and construction.

As always, Wenham spoke with reverent fascination of the natural world and the clues to be gleaned from it. For example, he described having seen a flock of spoonbills skimming low over the Nile. "Let one circumstance be marked—though they have fleeted past at a rate of near thirty miles per hour," he told his audience, "so little do they disturb the element in which they move that not a ripple of the placid bosom of the river, which they almost touch, has marked their track."[6]

Of course, Wenham shared his conviction that long, narrow wings lift more effectively than short, stubby ones. Aeronautical engineers call this relationship between a wing's length (span) and front-to-rear measurement (chord) its *aspect ratio.* Long, slender wings have higher aspect ratios than short, broad ones. If a particular wing's span is seven times its average chord, its aspect ratio is 7:1. Aspect ratio is in fact one of the most important parameters of wing design. Sailplanes generally have the highest aspect ratios and jet fighters the lowest. (Not all wings have parallel leading and trailing edges, of course; some wings are elliptical, and others, such as swept wings, taper as they extend outward. In such cases, the *average* chord length is used to calculate aspect ratio.)

The Aeronautical Society published "On Aerial Locomotion" in 1867. Lilienthal, Hargrave, Langley, and Chanute were among the flight researchers influenced by it. Hargrave particularly applauded

Wenham for his advocacy of kite experimentation and "superposed lifting surfaces" (multiplane wing configurations).

To further disseminate this Englishman's seminal ideas, Chanute paraphrased parts of his paper in the pages of *Progress in Flying Machines*. About the same time, a wealthy Bostonian named James Howard Means reprinted the entire paper in the 1895 edition of his *Aeronautical Annual*, a compendium of writings "devoted to the encouragement of experiment with aerial machines."[7]

The Wrights first became familiar with Wenham's ideas when Wilbur wrote to the Smithsonian Institution in 1899 asking for information about and a bibliography of flight. Igor Sikorsky was another likely beneficiary of Wenham's thinking; if so, it explains how Sikorsky knew to give his giant Grand and Ilya Muromets airplanes such high-aspect-ratio wings.

As stated, it bothered Francis Wenham that a dearth of knowledge had sabotaged his efforts to build a working glider. There had been too few answers and too much guesswork as he cobbled the craft together. By the time the pragmatic Englishman gave his Aeronautical Society lecture in June 1866, he had arrived at a decision of profound consequence. "I propose shortly," he informed those assembled, "to try a series of experiments by the aid of an artificial current of air of known strength, and to place the Society in possession of the results."[8]

The result of this public commitment would be the *wind tunnel*, a new aeronautical test device fundamentally more capable and accurate than Cayley's whirling arm. As its name suggests, a wind tunnel is a chamber through which air is ducted at a constant velocity so that the aerodynamic performance of different shapes can be assessed.

Collaborating with John Browning, who did the constructing, Wenham had the world's first wind tunnel up and running at the start of the 1870s. It was a magic chamber in which patient thinkers would coax flight's secrets out of hiding.

A hollow box about 10 feet (3 meters) long, the tunnel was open at the ends and employed a fan to drive air through at a constant rate.

A window let experimenters observe how different wing shapes performed in this artificial airflow.

The airfoil section under test was mounted on a balance that deflected under aerodynamic forces. By testing different airfoils at various *angles of attack* (orientations to the airflow), the lift and drag characteristics of these airfoils could be tabulated and plotted.

This first-ever wind tunnel provided very little usable data because it suffered from flaws. The air flowing through it was neither smooth nor constant, and the mounting balance was insufficiently sensitive. Horatio Phillips, a young member of the Aeronautical Society, built an improved tunnel in the 1880s that addressed these failings, and began pioneering research that continues today.

Cayley said that wings should be cambered, not flat. He based this guidance on methodical airfoil studies. He measured their lift, the movement of their center of pressure as the angle of attack changed, and their air resistance. He even correctly theorized that airfoils create lower pressure above the wing than below.

Having learned so much, it is ironic that he offered future researchers little specific guidance on the wing's shape. Moreover, his own full-size gliders relied on fabric wings indiscriminately billowed to a camber by the air. While he foresaw the need for rigid airfoils in the future, he also apparently felt such refinements were not necessary for the low speeds and weights of his gliders. Consequently, the craft he built late in life were paragliders.

In contrast, Otto Lilienthal's gliders had rigid wings. The great German pioneer chose an arc for his airfoil, whereas birds' wings employ parabolic cambers with the greatest curvature near the front. Efficiently honed by evolution, this latter shape affords the most lift.

It is just possible that Lilienthal was fooled in his anatomical studies. Absent active muscular control, a dead bird's wing might not always display its normal flight camber. However, it is more likely that Lilienthal fell prey to the ancient Greeks' fondness for classic geometric forms, a love strongly echoed in the teachings of nineteenth-century Europe. Whatever the reason, Lilienthal selected the arc—a section of a perfect circle—for his airfoil camber. This decision compromised the stability and performance of his gliders; it may even have cost him his life.

Down near the Baja border east of San Diego, twenty-five-year-old Californian John J. Montgomery and his younger brother James jumped down from the plank seat of their buckboard. Setting aside rifles, they hefted a compact glider off the wagon. The horse shied as they carried the two-winged contraption to the edge of a gradual hill nearly a mile long.

Setting the glider down, the young men surveyed the landscape. California's Otay Mesa spread before them, an undulating panorama of sagebrush, cactus, buckwheat, and sunflower. Home to rattlers, tarantulas, scorpions, foxes, coyotes, vultures, and other wildlife, its scrub-covered folds were also dotted here and there with ranches and somnolent cattle. A red-tailed hawk burst from a tall lemonadeberry and plunged to the ground, rising again with a lizard in its talons. Sunlight flashed russet in the bird's tail, and the world was again still.

It was August 28, 1883, a sweltering day in the American Old West. Less than two years had passed since the infamous shootout at the O.K. Corral in Tombstone, a lawless mining town in the neighboring Arizona Territory. One year had passed since notorious train robber Jesse James fell dead, shot from behind by a fellow outlaw.

Slightly built with dark hair and eyes, John Montgomery lifted his creation high. He slipped a leg over an underslung wooden keel padded like a saddle. Weighing 40 pounds (18 kilograms), this glider might have been too small for others, but Montgomery himself weighed just 130 pounds (59 kilograms). The craft had stubby wings and a fan-shaped tail like that hawk's. Sunshine on its taut cotton covering made him squint.

Montgomery's sister Jane had stitched the fabric back at the family ranch. Everything else he had built himself. For the wing's ribs he had chosen ash, a strong and springy wood favored by Native Americans for hunting bows. Steam-softening thin strips of this wood, he dried them in forms, lending them the curved contour he had selected for his artificial wings.

What's interesting is the wing shape Montgomery chose. It was a parabolic curve, perhaps the first ever used by a human being to fly. John Montgomery had learned the *right* lesson from studying birds.

Montgomery looked at his brother. James stood ready a dozen steps down the slope, a rope in his hand tied to the glider's front.

"Now!" John shouted, rushing forward.

James took off headlong down the slope, yanking hard on the rope to slingshot the glider into the air. Releasing the tether as instructed, he watched in awe as his brother sailed past in full flight. His elation turned to fear as the glider continued down the hill, then to relief as John touched down safely on his own two feet.

He ran to join his brother. Carrying the glider, they trudged back up the hill and paced off the distance as they went. John J. Montgomery had flown 600 feet (180 meters). "There was a little run and a jump and I found myself launched in the air," he later wrote. "A peculiar sensation came over me. The first feeling in placing myself at the mercy of the wind was that of fear. Immediately after came a feeling of security when I realized the solid support given by the wing-surface. And that support was of a very peculiar nature. There was a cushiony softness about it, yet it was firm. When I found the machine would follow any movement in the seat for balancing, I felt I was self-buoyant."[9]

Montgomery's may have been the first manned glider to fly in the Western Hemisphere. It would be another ten years before Otto Lilienthal took up gliding in Germany.

Born in 1858 in Yuba City, California, John Joseph Montgomery had come into the world aching to fly. His mother later recounted how as a toddler John lay on a pillow, flapping his arms and vividly imagining himself aloft. As Montgomery grew older, he built kites and savored the elemental connection they provided to the wind and sky. Large soaring birds, a prominent feature of the American West, fascinated him.

Montgomery was living in the San Francisco area when in July 1869, at age eleven, he attended the public testing of an unmanned flying machine by elderly northern California newspaper publisher

Frederick Marriott. Called the *Avitor Hermes Jr.* (*avitor* being a variant of the word *aviator*), this dirigible-airplane hybrid flew under the impulse of a 1-hp steam engine. The lumbering machine's sausage-shaped gasbag made it only slightly heavier than air, allowing it to fly.

Marriott had been interested in flight ever since his days a quarter century before as an officer of Great Britain's Henson Aerial Steam Carriage company. Yes, here was the same Frederick Marriott who had commissioned the evocative artwork showing the world what the airplane would someday be. While his post–Civil War *Avitor Hermes Jr.* was the first powered flying machine to fly in North America, however, it was also a technological dead end.

John Montgomery's flight of 1883 appears to have been the last of his youthful experiments. He earned a Ph.D. and pursued a teaching career in northern California. Dabbling in aviation over the passing years, he eventually resumed glider trials in the opening years of the twentieth century. Having lived to see the airplane invented, California's first aviation pioneer died in his early fifties in a 1911 glider crash.

Like so many aviation dreamers in the late nineteenth century, John Montgomery corresponded with Octave Chanute, who devoted space to Montgomery's work in his 1894 book *Progress in Flying Machines.* Although this coverage is brief and makes no mention of parabolic airfoils, the 1896 glider Chanute built with Augustus Herring's help likewise featured a parabolic camber. Consequently, when the Wrights adopted the Chanute-Herring glider as their starting point, they also inherited its bird-inspired airfoil as a point of departure.

While a parabolic camber was a step in the right direction, it was not the ultimate solution to the aerodynamic challenges of wings because the Chanute and Montgomery airfoils were based on guesswork. Mimicking the camber of birds' wings without understanding the underlying aerodynamic principles, they were far from optimal.

A great deal more thinking and learning needed to be done before the airplane would emerge. It called for a degree of scientific rigor never before witnessed in the field of flight.

The Wright 1900 Glider.

Wilbur and Orville tested their first glider at Kitty Hawk, North Carolina, in 1900. The brothers had intended to fly this machine, but its performance was so poor that they ended up using it primarily as a kite, sometimes loaded with sand or chains for ballast.

They measured its angle in flight and that of its rope in wind speeds ascertained by an anemometer. A grocer's scale told them how hard this kite pulled on the rope. From these observed and recorded results, they calculated its lift and drag.

The Wrights had sized this machine using Lilienthal's lift tables as republished by Octave Chanute. Based on that data, they had believed their first manned glider had sufficient wing area. Now direct experimentation told them it was generating only a little over a third of the expected lift. Something was very definitely wrong.

They returned the next year with a glider of almost twice the wing area. Although they logged many flights in this 1901 machine, it too performed poorly. At the end of the flying season, they went back to Dayton thoroughly discouraged, wondering if they should give up their pursuit of flight.

The Wrights had taken up gliding as a hobby to engage their minds and hands. It was initially Wilbur's interest, but Orville too had succumbed to flight's siren call. Talking things through in late 1901, they knew they couldn't let it go. But with Otto Lilienthal's lift tables now called into question, what were they to do?

There was only one answer: if Lilienthal's aerodynamic data were wrong, then they must work up accurate lift and drag tables of their own.

The trees in Dayton were golden brown. Fall crispness filled the air. In a room at the back of the Wright Bicycle Company, Wilbur cut out a small rectangle of thin sheet steel with tin snips. On a worktable, he hammered this flat piece into the airfoil curve he desired. He used solder and wax to build up the leading edge to the right thickness where necessary, and more solder to attach mounting prongs.

Wilbur then brought this completed airfoil to the wind tunnel he and Orville had constructed. Mounted waist high on sturdy legs, it was 6 feet (1.8 meters) long and square in cross section, 16 inches (40 cm) wide by high. At one end was a steel shroud with a fan. When turned on, the fan forced air through the tunnel at a constant 25 mph (40 km/h). The air passed through a metal honeycomb to smooth and align its flow before reaching the test chamber.

Wilbur and Orville Wright built a wind tunnel to probe flight's secrets at their bicycle shop in Dayton.

By 1901, there were perhaps ten wind tunnels in the world. None of them was being put to anything like the systematic, scientifically focused use the brothers were making of theirs. The sketches they made identifying the tiny forces to be measured, and the sensitive mechanical balances they devised to capture these readings, show how cleverly they rose to this challenge.

There were two of these mechanical balances. Made out of old hacksaw blades and bicycle-spoke wire, they worked by comparing astonishingly small forces. One balance assessed an airfoil's lift and the other its drag (the Wrights called the latter *drift*). From their results, the Wrights could also determine a given airfoil's *center of pressure*, the point at which the lift is concentrated.

Wilbur locked his newly completed airfoil section in one of the balances at the angle to be tested. Lowering the balance into the wind tunnel, he secured it to the bottom and closed the glass top. At his nod, Orville turned on the fan and stood stock-still because walking around the room could affect the airflow, throwing off the delicate test results.

The tunnel thrummed as air washed at high speed across the sideways-mounted airfoil. Its reaction in the slipstream deflected the needle of a gauge. Peering through the glass, Wilbur wrote down the reading.

To learn what they needed to know, the Wrights had started with preliminary tests of a dizzying variety of airfoils. They assessed wings of rectangular, triangular, elliptical, and circular planform; wings with different cambers and aspect ratios; wings by themselves (monoplane) or in combination (biplane, triplane, or tandem like Langley's aerodromes); and multiple wings with variations in the spacing between them to observe the effect of wing gap.

The brothers tested some two hundred airfoils over four weeks. Wilbur became so proficient at cutting and shaping airfoils that he could complete a new one in as little as fifteen minutes. Painstaking as it was to build up an entirely new body of knowledge, there was also the quiet thrill involved in exploring uncharted intellectual territory.

"We had taken up aeronautics merely as a sport," Orville later wrote. "We reluctantly entered upon the scientific side of it. But we soon found the work so fascinating that we were drawn into it deeper and deeper."[10]

Here was Kitty Hawk in microcosm. In fact, it was more productive than that because empirical results could be obtained without having to travel, construct full-scale gliders, or carry them again and again to the tops of sandy hills in long trudges. Better still, the brothers were spared nature's capriciousness because the gusting, quartering, and faltering sea breeze made accurate test results hard to come by at Kill Devil Hill.

As fall turned to winter, the Wrights were focusing their attention on upward of fifty wing shapes that looked particularly promising. Each was tested at angles of attack ranging from zero to 45 degrees in 2.5-degree increments, first in one balance and then in the other. Calculations made according to simple equations allowed them to convert this raw lift and drag data to usable results. The tabulated aerodynamic performance of the different airfoils could then be plotted and compared.

As their knowledge grew, the brothers saw why their 1900 and 1901 gliders had performed poorly. Smeaton's coefficient—a value named for British civil engineer and physicist John Smeaton, who in 1759 first measured the pressure exerted by moving air—was off by a considerable amount. As a result, Lilienthal's tables of calculated airfoil lift also had been off.

Even so, the Wrights' studies increased their admiration for Lilienthal, whose measurements of *relative* airfoil performance were surprisingly accurate for someone who tested with a whirling arm rather than a wind tunnel. By the same token, the Wrights found Samuel Langley's comparative lift data so flawed as to be worthless.

As this intensive testing drew to a close, the Wrights settled on one airfoil in particular that showed the best overall performance. Called Wing #12, it was a narrow rectangle with a 6:1 aspect ratio (it was six times longer in span than in chord). This parabolic airfoil had a camber just one-sixteenth as high as long.

The next glider, built in the new year, incorporated this longer, narrower wing with its flatter camber. Tested the following summer

Controllable with the addition of a movable rudder, the Wright 1902 Glider set the stage for the Wright brothers' invention of the airplane the following year.

and fall at Kitty Hawk, the Wright 1902 Glider matched its predicted performance and rewarded the brothers with the longest glides yet.

"Our new machine is a very great improvement over anything . . . anyone has built," Wilbur wrote in October 1902. "Everything is so much more satisfactory that we now believe that the flying problem is really nearing its solution."[11]

Enlarged further to accommodate an engine, this breakthrough wing became the aerodynamic and structural heart of the Wright 1903 Flyer, history's first airplane. Louis Blériot, Glenn Curtiss, Igor Sikorsky, and countless other early fliers would trust their lives to essentially similar wings for the next fifteen years.

Then something unexpected happened. Human beings bid farewell to the thin, bird-inspired wings that had first carried them aloft. It was all because a German university professor in his fifties asked a radical question.

6 WINGS, PART II

CLOUD-CUTTING CANTILEVERS

You cannot fly like an eagle with the wings of a wren.

—WILLIAM HENRY HUDSON (1841–1922)

Thin wings gave aviation its start. They did most of the fighting over Europe in World War I. And thanks to Hugo Junkers, they died a casualty of that conflict.

Junkers is, of course, the German industrial engineer we met earlier because of his all-metal airplanes. In an era when wings were thin and bird-like, he alone thought to wonder whether *thick* wings too could be made to fly. It was so heretical an idea that it is a bit startling he came up with it, but there was method to his madness.

The strut and wire bracings between the thin wings of World War I biplanes are fine from a structural standpoint. Aerodynamically, however, they come at a high price because they significantly retard the airplane's passage through the air. As a result, externally braced biplanes are known for high drag, not just high lift.

Early monoplanes didn't do much better on this front. The Fokker Eindeckers, Rumpler Taubes, and other World War I monoplanes

The Rumpler Taube shows off its thin, bird-inspired airfoil. Although Taube *is German for dove, Austrian designer Igo Etrich reportedly modeled this airplane's graceful wing after the zanonia seed.*

(like the Blériots and Antoinettes that preceded them) had fewer struts but featured correspondingly more wires. Extending outward from a kingpost atop the fuselage, they braced the thin wings from above while more wires from below tied them into a bottom post or perhaps the landing gear.

Physics decrees that aerodynamic drag increases proportionally to the square of an object's speed through the air. Thus, an airplane flying twice as fast encounters four times the drag. This physical limitation became all too apparent as people tried to go faster using World War I–era aviation technology. It's also why old-time pilots often joked that their airplanes came with built-in headwinds.

Although the wing struts of a biplane looked to be most of the problem, it turned out that bracing wires, thin as they were, were just as culpable. The cylindrical cross section of a wire presents a blunt, symmetrical, and unstreamlined profile to the relative wind that makes it a particularly bad shape to drag through the air. The result is turbulence that translates into drag.

One clue that this was happening was that bracing wires sometimes vibrated audibly in flight. When World War I flying aces spoke of hearing their wires singing during high-speed chases, they weren't merely waxing poetic; airflow ambivalence was vibrating these wires like a bow drawn across the strings of a great cello.

Biplane manufacturers finally addressed this drag penalty in the 1920s—and in the process ended the singing—by substituting "flying wires" of teardrop cross section for the piano wires of flight's earliest days. This helped a bit, but the real solution was to invent wings strong enough to dispense with external bracings. And for that, the wings needed to be thick.

Born in 1859, Hugo Junkers grew up in Rheydt, today an industrial suburb of Mönchengladbach in west-central Germany. Not interested in his family's textile business, Junkers took degrees in mechanical and electrical engineering. A hardworking entrepreneur, he helped design products and created companies to bring them to market. One of his firms manufactured diesel ship engines. Another produced steam boilers and water heaters that he had designed.

In 1906, Junkers traded the pressures of industry for a teaching post at Aachen, in his home state of North Rhine–Westphalia. Romans had walked Aachen's historic streets, and Charlemagne had made the town his capital. But the Aachen of Junkers' day looked to the future because of its famed technical university.

Two years after joining the university's staff, Hugo Junkers accidentally entered the field of aviation when a colleague asked him for engineering help with an airplane he was designing. Had that professor called on an established aviation expert, history would have taken a different turn. But instead he asked Junkers, a complete neophyte when it came to flying machines.

Junkers had little reverence for the paths others had taken. What's more, he knew and cared little about bird wings and had no wish to emulate them. Instead, at his age and being a forceful personality, he saw flight as a chance to apply hard-won knowledge to benefit a new field. It was a challenge he relished.

All of this predisposed Junkers to think in unconventional terms. Being experienced in steel fabrication, he immediately considered all-metal construction. Never mind that others had ruled out steel as being too heavy for more than sparing aviation use. And while Junkers didn't really seek to make a water heater fly, as some suggested, there was unquestionably boiler DNA in his Junkers J 1 of 1915, the first all-metal airplane to fly.[1]

Significantly, the structural engineer in Hugo Junkers couldn't help wondering whether thick wings would work. If the answer was yes, it meant that wings could be designed robust enough to carry all the loads of flight internally. If so, they would need to be supported only at one end, like the span of a cantilever bridge; there would be no need for external bracings.

Of course, such wings would have to be considerably thicker than any in existence. Consequently, when Junkers began his aerodynamics research in 1913–14, his wind tunnel studies focused on one burning question: could fat cambered airfoils sustain an airplane in flight the way thin ones do?

The answer turned out to be yes. In fact, as the world would soon discover, fat wings performed *better* than thin ones.

Thin airfoils are inherently dangerous. With their sharp nose profiles, they stall with little or no warning to the pilot.

An aerodynamic stall is what happens when the angle between the wing and the air it is passing through becomes too great for the airplane's speed, weight, and available power. The airflow separates from the wing, causing the airplane to quit flying and start falling.

To recover from a stall, the pilot ceases to pull back on the stick,

Junkers employees demonstrate the strength of the Junkers G.23's fully cantilevered wing in 1924.

allowing the airplane to regain flying speed and continue on its way after a loss of altitude. If the airplane falls off on one wing when the stall breaks, the pilot may also have to counteract a spin by applying opposite rudder as part of the stall recovery.

All of this is simple enough *if* you have enough altitude to complete a successful recovery when the airplane stalls. If not, you're out of luck. That's why advance warning of incipient stalls is so important.

As it turns out, thick wings with their blunt noses do provide this warning. Instead of the airflow separating all at once at the front of the wing, aerodynamic separation at high angles of attack on thick airfoils begins aft and progresses forward. This causes the wing to buffet as the slipstream becomes increasingly turbulent. Sensing this buffeting through the controls, the pilot knows to reduce pitch angle, increase power, or both to avoid stalling.

Safety is important, but that alone did not drive the world to thick wings. Instead it was greater speed, which could be obtained only with aerodynamically clean wings.

World War I was raging and the Central Powers needed more weapons. Germany's wartime authorities contracted with Junkers for airplanes but worried about his lack of experience building flying machines. To help him come up to speed on the manufacturing front, they brought in Anthony Fokker.

Nicknamed *der fliegende Holländer* (the Flying Dutchman), Anthony Fokker was the dashing young man at the helm of Germany's most productive airplane company. He earned lasting fame during the war by building the all-red triplane in which Germany's "Red Baron"—top ace Manfred von Richthofen—scored many of his eighty aerial victories. A clever inventor, Fokker also revolutionized aerial warfare by inventing the synchronizing gear that allowed machine guns to fire through spinning propellers without hitting the blades.

Elderly Junkers had little use for Fokker, the twentysomething wunderkind. Fokker likewise chafed under the forced collaboration and was anxious to get back to his own company. Both men went their separate ways as quickly as possible.

From that brief association, however, Fokker took away Junkers' utterly invaluable idea of thick wings. His company would apply this knowledge to three of its fighters: the Dr.I triplane, the D.VII biplane, and the D.VIII (E.V) monoplane. The best fighter plane of the war, the Fokker D.VII was also technologically the most influential after hostilities ceased. Its formula would sire a line of successful commercial airliners between the world wars. These would fly in many nations, including the United States.

Knute Rockne was head football coach at the University of Notre Dame some 90 miles (145 kilometers) south of Chicago. In a dozen thrilling years, Rockne led his college team to 105 victories with just 12 losses and 5 ties. By 1931—fresh off a fifth undefeated season and sixth national championship—he was hands down American football's most famous coach.

On March 31, 1931, the forty-three-year-old Norwegian American waited to board a Transcontinental and Western Air (TWA) flight at Kansas City, Missouri, where he had visited his sons. He was headed for Los Angeles, having been summoned to Hollywood to participate in the production of *The Spirit of Notre Dame*, a feature film based on his team's exploits.

Located along the Missouri River, the Kansas City airport was an open field with a few Spartan buildings. Americans traveled by train back then, and the nation's fledgling air carriers catered only to the wealthy and those who placed a premium on speed. Hollywood did, and so the studio had sent Rockne a plane ticket to Los Angeles via Wichita and points west.

The flight arrived. Knute Rockne and his fellow travelers boarded the Fokker F-10A, a high-wing trimotor with seating for a dozen passengers. The engines sputtered to life and the plane trundled to the far end of the field. Turning into the wind, it roared down the grass and clambered into the air.

For more than an hour the Fokker flew steadily southwestward. Suddenly a wing broke away. The stricken craft plunged to a Kansas wheat field below, killing all aboard.

Knute Rockne's name was a household word across America. His values—humility, generosity, and a commitment to hard work—were universally admired. The national outcry over his untimely death reverberated in the media, lending urgency to the accident investigation. Speculation initially focused on an encounter with a violent squall. However, no thunderstorms had been reported in the area. Then as the wreckage was examined a more sinister picture emerged.

Like all Fokkers, the F-10A had a sealed plywood-skinned wing. As it turned out, cracks had developed in the crashed machine's wing. Rainwater had seeped in, softening the plywood and dissolving organic glues. Like a ticking time bomb, this undetected deterioration had progressed to the point where the weakened structure simply could take no more. Although the airplane was just eighteen months old, it had failed catastrophically under normal flight loads.

All Fokkers in the U.S. fleet were summarily grounded and man-

Fallout from the 1931 crash of a Fokker F-10A, which killed football coach Knute Rockne, hastened U.S. adoption of all-metal airliners.

datory inspections were performed. When similar deterioration was found in other Fokkers, trust evaporated in wooden flight structures (on Fokkers, only the wings were wooden). All of this came as a great blow to Anthony Fokker, whose F.VII trimotor in particular had found wide use in the U.S. civil fleet.

Tony Fokker had immigrated to the United States, become a naturalized citizen, and started two airplane factories in New Jersey and West Virginia. His airplanes had set many international records and were the state of the art. But the Kansas crash—together with the spreading Great Depression, triggered by the U.S. stock market crash of 1929—effectively put him out of business.

Rockne's loss thus became a catalyst that propelled aviation forward. It ensured that subsequent U.S. airliners would be built entirely of metal, wings included.

On May 20–21, 1927, a slender twenty-five-year-old U.S. airmail pilot named Charles Lindbergh flew his monoplane *Spirit of St. Louis* all the way from New York to Paris nonstop. His thirty-three-and-a-half-hour solo flight thrilled the world.

Photographs suggest one reason why Lindbergh's flight had such cultural resonance. Hollywood's finest casting and prop departments could not have matched the visual perfection of this clear-eyed young Swedish American flier and his graceful silver ship. But more was at work here than poetic appeal.

By directly linking two of the world's great cities, Lindbergh's flight audaciously suggested commercial air travel might someday be possible between the world's continents. Airplanes would not just rival the train; someday they might also give ocean liners a run for their money.

With Lindbergh, America seemed to become "air-minded" overnight, though in fact technological progress had been ongoing. A critical mass had been reached, however. Assuming leadership in the spectrum of flight-related technologies, the United States took off and left the rest of the world behind.

The Wright brothers had single-handedly dominated aviation through the year 1908, after which they quickly fell behind.

Charles Lindbergh and the Ryan NYP Spirit of St. Louis, *1927.*

France assumed the lead and held on to it through most of World War I. Before that war ended, however, Germany had claimed the mantle of aviation leadership. Dominant in the sciences since the latter part of the nineteenth century, German inventors, mathematicians, and theoreticians remained in the forefront throughout the 1920s.

Now with the 1930s, America's turn had come again. Serving notice that this was the case were two single-engine U.S. mail planes that took to the skies at the start of the decade. The first was Jack Northrop's all-metal Alpha, which flew in March 1930 in the Los Angeles area. The Alpha—the first production airplane to employ all-metal semi-monocoque construction—had a fully cantilevered wing unsullied by struts or wires.

Some two months later and 950 miles (1,500 kilometers) up the West Coast of North America, Seattle saw the first flight of the Boe-

Returning by ship to America, Lindbergh and his Spirit of St. Louis *made a triumphal 1927–28 tour of the United States and Latin America, drawing large crowds wherever they landed.*

ing 200 Monomail. A rival airplane built to the same configuration, the Monomail was Boeing's first stressed-skin airplane. The company elected not to produce the Monomail because it had grander plans on the commercial front.

In January 1933, Adolf Hitler became chancellor of Germany. Across the Atlantic, Franklin Delano Roosevelt, Hitler's antithesis, took office in March as president of the United States. That year in New Deal America, the chocolate-chip cookie was invented, Prohibition was repealed, and history's first two modern commercial airliners took to the skies.

The first modern passenger airliner, the sleek and fast Boeing 247, shaved eight hours off the transcontinental flight time of the era's Fords and Fokkers. The 247 carried ten passengers and a crew of three plus baggage and 400 pounds (180 kilograms) of airmail or freight. In an era of cumbersome, ungainly airliners—including Boeing's own fabric-covered Model 80A trimotor biplane—it crossed the United

States in twenty-one hours westbound and just nineteen flying east with the prevailing winds.

The Boeing 247 was an all-metal semi-monocoque monoplane with a fully cantilevered low wing and two engines. It also featured a retractable landing gear, an autopilot to reduce pilot workload, gyroscopic instruments for night and bad-weather operations, and deicer boots to shed ice from its wings and tail in flight.

The structural heart of this astonishing transport plane was a wing with built-up metal spars reinforced by bridge-like diagonal members in a configuration known as the Warren truss. Riveted aluminum sheeting surrounded this structure, locking it together and sharing the structural loads. The resulting wing was clean, light, and strong.

In a bold move to update its entire fleet, United Aircraft and Transport Corporation (today United Airlines) ordered fifty-nine Boeing 247s. This tied up at least the first year's production, forcing United's competitors to look elsewhere.

Needing to replace its discredited Fokkers and dowdy Fords, Transcontinental and Western Air solicited bids from five companies for a new high-performance trimotor. Douglas Aircraft of Santa Monica, California, accepted the challenge but proposed instead a twin-engine design more advanced than Boeing's.

Even as the Boeing 247 flew in February 1933, Douglas was building this all-metal semi-monocoque airliner. The twelve-passenger DC-1 (*DC* for "Douglas Commercial") flew that July and went on to set many records. Before placing this design into production in 1934, however, Douglas stretched its fuselage to accommodate one more row of seats. The result was the DC-2, which carried fourteen passengers.

Although first to market, Boeing's 247 found only limited commercial success because it simply carried too little payload to exploit the potential of its new technology. The Douglas DC-2 did somewhat better, offering incrementally greater speed, comfort, and range as well as 40 percent more passenger capacity. Moreover, the DC-2 offered the advantages of a higher *wing loading*.

Wing loading is the loaded weight of the airplane divided by the area of its wings. An important design parameter, it largely determines an airplane's capabilities. Those with low wing loadings (for example, old biplanes and modern light airplanes) need less installed power to achieve flight and operate at lower speeds. In contrast, high-wing-loading designs (jetliners and jet fighters, for instance) require more power and fly much faster.

Igor Sikorsky's Ilya Muromets of 1914 provides a good example. It had the wingspan of a World War II Boeing B-17 bomber and roughly the equivalent wing area. However, the Sikorsky also had just one-eighth the engine power of a B-17. Consequently, the Ilya Muromets needed a much lower wing loading to achieve flight. With a gross weight one-fifth that of the B-17, the Ilya Muromets also had about one-fifth the wing loading, which greatly limited its speed and payload. This comparison illustrates why the best World War I airplanes pale in comparison to those of World War II, and why today's high-performance military machines in turn far outstrip those of that latter conflict.

A key benefit of high wing loadings is that the wing is less affected by turbulence in the air through which it is passing, so passengers enjoy a smoother ride. On the minus side, high-wing-loading airplanes need longer runways because they take off and land at higher speeds. This is the reason for wing flaps, as described below.

Boeing designed the Model 247 with a wing loading only slightly higher than that of the airplanes it replaced, so it needed no wing flaps. In contrast, Douglas audaciously jumped to a wing loading half again higher than the 247's, giving the DC-1 and its successors superior performance. However, this design choice also meant that these Douglas airplanes required flaps to keep their takeoff and landing speeds sufficiently low.

Wing flaps are movable parts of a wing that angle down to give it greater camber, increasing both its lift and aerodynamic drag. Flap extension converts a high-speed wing into one suited to a lower operating-speed range. With flaps fully down, an airplane cannot fly very fast, but it can land more slowly and use smaller airports with shorter runways.

There are many flap configurations, the most significant being the *plain flap*, *split flap*, and *Fowler flap*. A plain flap is a hinged portion of the wing trailing edge located between the ailerons and wing roots. When actuated, a plain flap rotates down to give the wing greater camber. A split flap resembles a plain flap except that it is tucked beneath a fixed trailing edge. When actuated, this underside flap diverges from the wing's top like an opening book cover.

The most aerodynamically efficient system is the Fowler flap, which is significantly more sophisticated than hinged flaps. Fowler flaps extend aft and downward, not just downward, to increase the wing's total area, not just its camber. Fowler flaps have been around since the late 1930s. The Messerschmitt Bf 109 had them in the Spanish Civil War and World War II. The Lockheed P-38 Lightning and Northrop P-61 Black Widow also had them in the latter conflict. In general, however, Fowler flaps were not widely used until the jet age.

Many civil and military jet transports have the *slotted flap*, which is the Fowler flap with an opening between the flap and the wing. This allows some of the high-energy air beneath the wing to flow through the slot and stream down over the top of the flap. This blast reduces drag and lowers the airplane's stall speed by keeping the airflow attached to the flap longer (aerodynamicists call this "reenergizing the boundary layer").[2]

Jetliners have Fowler flaps with one, two, or even three segments (these are termed single-, double-, or triple-slotted flaps). While such flap systems are aerodynamically very efficient, they are also very expensive to build and maintain. In recent times, designers have learned how to give jets equivalent efficiency with greater simplicity.

Modern jet transports also have *leading-edge devices* that further convert their high-speed wings into low-speed wings. *Slats* are the most common leading-edge devices. A slat is a small leading-edge flap that slides forward and downward to further increase the wing's total area and camber (in addition to Fowler flaps, the Messerschmitt Bf 109's advanced wing also had slats that deployed automatically at low speed).

Some jetliners have an alternative leading-edge device called the *Krueger flap*. A Krueger flap is a hinged leading edge that rotates forward and downward into the slipstream to achieve the same aerodynamic benefits in a different way. When retracted, slats and Krueger flaps *are* the leading edge for the portion of the wing that they cover.

On your next commercial flight, look out the window at your jet's wing as it begins letting down to a landing. Depending on whether you're sitting ahead of the wing or behind it in the cabin, you'll get to see either the leading-edge devices or the trailing-edge flaps in action.

If your view is from behind, you will notice the flaps initially extending just a short distance aft when your airplane is still quite high. As it continues to descend and the crew commands additional increments of deployment, you will see the flaps travel farther aft and then begin angling down. All this added drag slows the airplane, reduces its stall speed, and increases the angle at which it descends to the runway, improving the pilots' view.

If instead you're watching from forward of the wing, you'll notice that extension of the leading-edge devices happens relatively late. By then, the jet is on a stabilized final approach to the runway. With all this going on, the wing looks dramatically different at touchdown. That doesn't last, though, because the crew retracts the flaps while taxiing to the terminal.

You may also notice a bit of flap being used on takeoff. The first increment of flap deployment increases the wing's lift without significantly increasing its drag, which helps the jet take off sooner.

Others had experimented with flaps, but the Douglas DC-1 marked the first time in history that an airplane was designed that *needed* flaps. Showing that the time was right, another design taking shape in Connecticut also employed a higher wing loading for increased performance. This was the four-engine Sikorsky S-42, Pan Am's first oceangoing flying boat, which would make its first flight in 1934. To keep its takeoff and landing speeds within reason, Sikorsky had given it a large plain flap that spanned almost the full length of its pylon-mounted wing.

For their new-technology airliner, Douglas engineers selected a split flap. They put flaps not just on the wings but also on the airplane's flat underbelly between the wings. And what a wing it was.

The heart of the DC-1 and its successors was Jack Northrop's *multi-cellular wing*, which he developed for his Alpha mail plane. Building further on the ideas of Germany's Adolf Rohrbach, this remarkable design locked together spanwise metal spars, stiffening members, and the enveloping skin to compartmentalize the wing internally into a series of mutually reinforcing boxes or cells. The result was enormous strength.

The Douglas DC-2 debuted in U.S. service with TWA in 1934. That same year, Dutch carrier KLM inaugurated European DC-2 services, putting European designs to shame. Few people imagined there could be anything better than the DC-2.

Then right away, almost by accident, the Douglas DC-3 came into being.

Trains were the paradigm for domestic air travel before World War II. In an era of slow airplanes, it was only natural for airlines to plan railroad-style sleeper services of their own. As darkness fell, passengers could then convert their seats into beds or retire into fold-down berths and doze off to the reassuring drone of engines.

With this in mind, American Airways (today American Airlines) and Eastern Air Transport each ordered Curtiss T-32 Condor II airliners for overnight sleeper services. America's last biplane passenger transport, the Condor II combined a fat fuselage with sturdy wings braced by struts and wires.

The Condor II was a product of the Curtiss-Wright Corporation, then the nation's largest aviation manufacturer. Having equipped the Condor with retractable wheels and aerodynamic engine cowlings, the company touted it as a "high-speed airplane," although it was derived from a 1920s bomber and evinced World War I–era technology. One can imagine the dismay at Curtiss-Wright when, a week to the day after the Condor II's first flight, the Boeing 247 took to the air. Overnight that event transformed Curtiss-Wright's Condor into a pterodactyl.

Cyrus R. Smith, the young businessman recently named president of American Airways, decided that his airline needed a new-

Engine cowlings and retractable wheels could not help the Curtiss T-32 Condor, a 1933 airliner rendered instantly obsolete by the Boeing 247.

technology replacement for the Condor, which cruised at 105 mph (170 km/h). Boeing's 247 cruised at 180 mph (290 km/h), but it was too small and narrow for Pullman berths. In any event, the Seattle company's production lines were all tied up for at least a year.

That left Douglas Aircraft in Santa Monica, California, to provide Smith's sleeper airplanes. The DC-2 also was too narrow for berths, but Smith had a solution to propose: redesign the DC-2 with a wider fuselage. As American envisioned it, the airplane would carry fourteen passengers as a sleeper or twenty-one passengers in day-plane configuration.

Widening an existing airplane is a *very* expensive proposition—it essentially means a new airplane—and Douglas, which was already selling as many airplanes as it could build, reasonably said no. However, Smith remained adamant, promising large orders and going so far as to secure external financing. Swayed by these inducements, Douglas finally agreed.

Douglas engineers gave this model a rounder fuselage that was wider and a bit longer than that of the DC-2. They also increased the wing area, gross weight, and engine power. The result was the Douglas Sleeper Transport, or DST, which in a day-plane layout was called the DC-3.

The Douglas DC-3 first flew on December 17, 1935, thirty-two years to the day after the Wright brothers succeeded at Kitty Hawk. It went on to change the world by offering a magic combination of payload, performance, and operating economy. Human beings had created an airliner that let its operators make solid profits even without airmail or other subsidies. No other airliner in history has ever been as dominant as the DC-3, which put the fledgling airline industry firmly on its feet.

With the coming of World War II, Douglas suspended commercial production and revised the DC-3 for wartime use. The U.S. military services called these twin-engine workhorses the C-47, C-53, or R4D. British and Commonwealth forces called theirs the Dakota, or less formally the Dak or Gooney Bird. Flying in all parts of the world, from Arctic tundra to tropical airstrips, these rugged airplanes proved crucial to the war effort. Without them, the United States and its allies could not have staged logistically around the globe.

World War II had two effects on commercial aviation. The first was the construction of countless airfields across the world, which laid down a global infrastructure for commercial flying when it resumed after the war. Flying-boat airliners, which pioneered transatlantic and transpacific flying in the 1930s, were left with no real role to play in the postwar world.

By accelerating the development of flight technologies, World War II also gave rise to a new generation of four-engine propeller transports. Among these great ocean-spanning piston airliners of the latter 1940s and 1950s were the Douglas DC-4, DC-6, and DC-7; Lockheed Constellation series; and Boeing 377 Stratocruiser. Many still fly in far-flung corners of the globe, although generally as cargo planes rather than passenger airliners.

Also still in the air are hundreds of DC-3s. Although the vast majority of these airplanes were built as military transports and are thus not true DC-3s, that famous airliner's designation is broadly applied to all surviving examples. Barring corrosion or damage, the wings of these robust workhorses never grow weary.

Introduced in 1956, the DC-7C "Seven Seas" was the last Douglas piston airliner.

With the structural challenges of wings solved, the next major leap forward came on the aerodynamics front. This was the development of swept wings for high-speed flight. Together with jet engines, wing sweep would further revolutionize aviation.

During World War II, some fighter pilots encountered inexplicable instabilities during all-out dives at full power that took the propeller fighter plane to or beyond its redline (maximum permissible) airspeed. As the airplane approached 70 percent the speed of sound, the airflow being accelerated over the tops of the wings and horizontal tail surfaces became locally supersonic. Shock waves formed that pushed the center of pressure farther aft, adversely affecting overall aerodynamics.

From the pilot's perspective the results were absolutely spooky. Seemingly with a mind of its own, the fighter tucked under into a tight dive, its control stick locked in a vise. Unable to pull out, many pilots lost their lives, although some managed high-speed bailouts while others successfully recovered at lower altitudes, where denser air moderated the troubles. In a few instances, pilots with nothing left to lose—perhaps sensing something through their controls—

experimentally pushed forward on the stick and were astonished to find themselves pop out of the dive. This particular manifestation was termed control reversal.

The culprit was compressibility (also called Mach phenomena), which cropped up as human beings began flying so fast that the air could no longer get out of the way in time. Instead of maintaining a constant density around the airplane, air molecules bunched up until shock waves formed, setting loose these compressibility gremlins.

There was obviously more to learn.

Aviation's early pioneers never knew *how* a wing achieves flight. They also didn't care. The simple fact that wings worked meant people could fly by following nature's lead.

In this sense, engineering, not science, gave us the gift of wings. Pascal, Bernoulli, Euler, and others laid down a body of theory to help explain the behavior of fluids, air included, at rest or in motion. However, that knowledge did not play a role in solving flight's challenges. Instead, science's role was initially to explain the underlying physics of flight after the fact.

It took a while to understand how wings actually work in terms of the circulation of air. Early in the twentieth century, the towering intellect in this cerebral quest was Dr. Ludwig Prandtl, a physicist and applied mathematician hailed as the "father of modern aerodynamics." Born in Bavaria in 1875, Prandtl mathematically described fundamental fluid-flow principles, identified key aerodynamic concepts (including the boundary layer) that underpin our understanding of flight, and gave airplane designers the ability to begin predicting the performance of wings that had not yet been built.

Prandtl spent most of his career at Germany's famous Göttingen University, which for decades he and his graduate students kept at the forefront of global aerodynamics research. Many of these disciples went elsewhere, carrying with them the light of theory to advance humanity's conceptual understanding of flight.

Max Munk was one of them. Immigrating to the United States after World War I, Munk joined the National Advisory Committee for Aeronautics (NACA), predecessor to today's National Aeronautics and

Space Administration (NASA). At the start of the 1920s, Dr. Munk came up with the *variable-density wind tunnel*, which by employing higher air pressures allows accurate aerodynamic data to be obtained from the testing of subscale models. No longer would full-size airplanes have to be tested.

Hungarian aerodynamicist Theodore von Kármán was another noted Prandtl disciple. Immigrating to the United States in 1930, Dr. von Kármán directed the California Institute of Technology's Guggenheim Aeronautical Laboratory, cofounded the Jet Propulsion Laboratory, and would for many decades serve as a leading aerospace advisor to the U.S. government.

Long before there were high-speed airplanes, Ludwig Prandtl's mathematical musings gave the world a theoretical basis for grappling with compressibility and shock waves. Other scientists also contributed to the emerging body of supersonic theory. In the 1920s, Swiss scientist Jakob Ackeret—another Prandtl alumnus—reduced these individuals' complex theories into a simple method for calculating the lift and drag of supersonic airfoils.

In a landmark paper presented in Rome in 1935, Adolf Busemann, also a Prandtl student, became the first person to propose sweeping back an airplane's wings as a way to reduce shock wave formation in high-speed flight. Since it takes energy to generate shock waves, which are like a boat's wake but in three dimensions, sweeping the wings back promised to reduce the amount of power required to fly at high speeds.

Wing sweep delays the onset of Mach phenomena such as those that fighter planes sometimes encountered in World War II. Thus, sweeping the wings allows airplanes to cruise closer to the speed of sound, or at higher Mach numbers, before running into compressibility issues. Named for Czech-Austrian physicist Ernst Mach, *Mach number* refers to the ratio of a given airspeed to that of sound, which varies with altitude.

During World War II, the Messerschmitt Me 262—the world's first operational jet fighter—was 100 mph (160 km/h) faster than the best Allied propeller fighters. It had modestly swept wings, but evidence suggests that Adolf Busemann's insights were not responsible for this design decision. Instead, its wings angled back to keep the airplane in balance after different engines had to be substituted during its development.

Adolf Busemann's 1935 speculations in Rome were not yet broadly appreciated even a decade later. But as the war in Europe drew to an end in the spring of 1945, Allied technical intelligence teams advancing through Hitler's collapsing Third Reich came across wind-tunnel and other research data that greatly excited U.S. experts. Thanks to these finds, two American jet airplanes then in development—the North American F-86 Sabre and the Boeing B-47 Stratojet—were redesigned in midstream to incorporate swept wings.

Coincidentally in 1945, NACA aerodynamicist Robert T. Jones independently rediscovered the value of wing sweep. The elegant mathematical understanding of its benefits that Jones provided went beyond what Busemann had come up with. All at once, wing sweep was firmly at the forefront of human thought.

As World War II raged, the U.S. Army Air Forces embarked on high-speed flight research in partnership with fighter-plane manufacturer Bell Aircraft in New York State. This critical research program was well under way before the benefits of wing sweep became known.

Not wanting to lose time, and fearful of introducing a potentially confusing new variable, the USAAF—which became the separate U.S. Air Force in September 1947—did not have Bell redesign its rocket-powered X-1 research plane with swept wings. Consequently, when USAF fighter ace and test pilot Captain Charles E. "Chuck" Yeager broke the sound barrier on October 14, 1947, he did so with straight wings.

Shortly before the X-1's historic flight, North American Aviation in Los Angeles flew its F-86 Sabre, the first U.S. swept-wing jet fighter, which could exceed the sound barrier in a dive. And not long after Yeager's success, the Boeing B-47 Stratojet jet bomber flew in Seattle on the forty-fourth anniversary of the Wright brothers' flight. An arrow to the future, the high-subsonic B-47 was the world's first large production airplane with swept wings.

On October 14, 1947, USAF Capt. Chuck Yeager broke the sound barrier in the rocket-powered Bell X-1.

The *delta wing*, another significant application of wing sweep, features a triangular planform that keeps the wing entirely within a supersonic airplane's shock cone for less drag. Initially proposed by German aerodynamicist Alexander Lippisch, delta-wing airplanes are generally semi-tailless because the aft-mounted wing also acts as the horizontal tail. Many American and French jet fighters have had delta wings, as have British bombers. The Concorde supersonic transport is also a delta-wing aircraft.

Fittingly, it is in commercial air travel—by far the airplane's most important role—that wing sweep has yielded the greatest benefits. Boeing gave its breakthrough 707 airliner the same 35 degrees of wing sweep as the B-47, although the two designs are otherwise entirely different. Douglas was a bit more conservative with its first jet, the DC-8 of 1959, which had 30 degrees of sweep. To this day, the subsonic jetliner with the greatest amount of wing sweep is the Boeing 747, with 37.5 degrees. The world's first commercial jet transport, the de Havil-

The revolutionary Boeing B-47 Stratojet took flight on the forty-fourth anniversary of the Wright brothers' success at Kitty Hawk.

land Comet of 1952, had only modest wing sweep and thus could not match the performance of the game-changing 707 and subsequent designs.

In addition to allowing jetliners to fly faster and use less fuel, wing sweep imparts greater overall stability. However, swept-wing aircraft are subject to a constant mild corkscrewing motion called Dutch roll. Named for the natural back-and-forth body swings that a Dutch skater might describe when crossing the ice, this coupled rolling oscillation is benign but can make passengers queasy. Consequently, all jetliners have yaw dampers that automatically negate this Dutch roll for a more comfortable ride.[3]

When flying commercially, you may notice that your jet's wings flex a bit during flight. This is entirely normal. Indeed, few transport planes have ever been designed with totally rigid wings. This flexing occurs only up and down because jetliner wings are entirely rigid torsionally. Of course, they are all enormously strong because government design certification requirements specify that they must be able to withstand loads half again greater than any they might ever conceivably encounter during a lifetime of operation. In fact, our mechanical wings whisk enormous loads at astonishing speeds and altitudes with by far the greatest safety ever achieved by any mode of mass transportation.

7 EMPENNAGE

WHALE FLUKES AND ARROW FEATHERS

Give me a place to stand and a lever long enough, and I will move the world.

—ARCHIMEDES (287–212 BCE)

Wellwood Beall had a dream. He was going to make history with a flying boat of unprecedented size, performance, and luxury. It would be the biggest and best yet of the ocean-spanning, boat-hulled airliners flown by Pan American Airways.

Beall, an aeronautical engineer from Colorado, worked for Boeing in Seattle. Although just twenty-nine years old in 1936, he was one of its ablest salesmen. It bothered him that his company, citing existing commitments and insufficient funds, had declined Pan Am's invitation to develop a new airliner. Working at home on his dining room table, he sketched out a huge new flying boat with lines like a whale. As he conceived it, this long-range airliner would have two decks, the upper for crew and baggage and the lower—configured as the aerial equivalent of an ocean liner—for passengers.

Beall's vision was persuasive and his colleagues changed their minds. Boeing would build Pan Am's new flying boat after all.

The prototype Boeing 314 Clipper rocked gently in the swells of Elliott Bay. Evening sunlight raked the Seattle skyline. A bracing west wind had died down.

It was June 7, 1938. Taxi tests completed, it was time to take the new airplane aloft for the first time. Wellwood Beall stood on the mooring barge and helped Boeing chief test pilot Eddie Allen and his crew hop across to the bobbing Clipper. Workers untied mooring ropes and the airplane drifted away, its engines coming to life one after the other.

Aboard the Clipper, Allen led his crew through preflight checks and engine run-ups. All was ready, and Allen—an aeronautical engineer with a master's touch on the controls—opened the throttles wide. Four radial engines bellowed as Boeing's Clipper accelerated, its planing hull lifting to skim the water's surface. Allen pulled back on the control wheel and the ship rose into the air.

Unbidden, the 314 dipped to the right. Allen tried to counteract the bank with his controls, but it was as if somebody else were flying the machine. He also found it directionally unstable, its nose unable to hold a steady course.

The wide right-hand arc brought them inland. Using the throttles, Allen lifted the low wing. Differential engine power provided enough control to set down after thirty-eight minutes aloft on Lake Washington, which offered calmer waters on the other side of Seattle. The crew heaved a collective sigh of relief as the machine decelerated, its hull settling into the water.

Eddie Allen knew what the problem was. The prototype 314's single vertical stabilizer and rudder were ineffective. Boeing replaced that single tail with twin stabilizers and rudders at the tips of the horizontal tail. When Allen flew the airplane again, stability and control were much improved but still inadequate. Boeing finally reinstated the center vertical fin and the Model 314 at last flew well.

Pan Am's Boeing 314s entered service in 1939 to inaugurate perhaps the most romantic and fondly remembered air travel of all time. Spanning 152 feet (46 meters) and weighing 41 tons (37 tonnes) at takeoff, Boeing 314A Clippers carried up to forty premium-fare passengers in overnight sleeping services and flew up to 3,500 miles (5,600 kilometers) nonstop.

An airplane's *empennage*, or tail assembly, is the rear fuselage together with its protruding vertical and horizontal stabilizers with their control surfaces. Connected by hinges to the rear of the stabilizers, these control surfaces deflect in the slipstream to alter the airplane's flight path. The one attached to the vertical stabilizer is the rudder, and the ones connected to the left and right horizontal stabilizers are the elevators.

Aviation decided quite early what an airplane's rear end should be and do. There have been variations on the above theme, and many outright exceptions starting with the Wright Flyer, but the vast majority of airplanes ever built have had empennages designed to this formula.

Sir George Cayley gave us most of it right off the bat more than two centuries ago. Aviation's Isaac Newton decided that an airplane's back end should have both horizontal and vertical surfaces, not just horizontal ones like a bird. He also said that they should help the airplane fly smoothly through the air (stability) and adjust its course as need be (controllability).

Where did Cayley get these ideas? For stability, the feathers of an arrow were an obvious inspiration. Since prehistoric times, fletchers have known that stiff feathers at the rear help an arrow fly true. They have also known that these feathers must be properly aligned for stable and thus accurate flight. The French term *empennage*, which literally means "addition of quill feathers," formalizes this association between arrows and aviation.

On the controllability front, Cayley drew inspiration from nature and human practice alike. Whales plunge deep by angling their great flukes downward and come up again by angling their horizontal tail upward. Fish too demonstrate a mastery of hydrodynamic control by aft-body deflection although their tailfins are vertical. Water being a dense medium, even the human act of swimming imparts intuitive insights into the cause-and-effect relationship between body motion and direction traveled.

Humans got the message early on. Boats equipped with steering oars or tillers at the rear date back more than two millennia. The idea that air too is a fluid medium came later, a product of Renais-

The delightful Santos-Dumont Demoiselle of 1909 featured a Cayley-style cruciform tail.

sance thinking and the scientific revolution. All of this made it conceptually easy for Cayley to translate hydrodynamic lessons to the emerging arena of aerodynamics. In fact, the idea of using rudders on airplanes leapt instantly to the minds of all early aerial experimenters. Reflecting this appropriation of maritime technology, the trailing control surface on an airplane's vertical stabilizer is called the rudder to this day.

Cayley did not actually distinguish between stabilizers and control surfaces. Instead, he proposed a cruciform tail with fixed horizontal and vertical surfaces that served as both stabilizers and control surfaces. This was possible because he mounted the entire tail on a universal joint, allowing it to deflect left, right, up, down, or any combination thereof as a single integral unit.

This idea actually works, as Alberto Santos-Dumont demonstrated in his diminutive Demoiselle monoplane of 1909. The world's first practical light airplane, the Demoiselle sported a Cayley-style movable tail. Since that time, however, aviation has steered clear of full-flying tails because of drawbacks that include poor aerodynamic performance.

Simple physics endorses Cayley's thinking. Stabilizers, which guide the airflow, can be located anywhere on the airplane. The Wright flyers initially didn't have vertical stabilizers; when they did, starting in 1905, the Wrights placed them at front and called them *blinkers* because they were reminiscent of the devices that let horses look only forward.

In fact, vertical and horizontal stabilizers actually work best at the rear, as do elevators and rudders. What's more, placing these stabilizers and control surfaces as far aft as possible allows them to be smaller, lighter, and contribute less aerodynamic drag than if they were located elsewhere on the airframe. This enhanced efficiency is why most airplanes have these components at the back end. It's also why they have their ailerons out near the tips of the wings instead of in at the roots. Intuitively we understand that the farther outboard we place these control surfaces, the more effectively they can bank the machine to the left or right.

What we're sensing is the mechanical advantage provided by a longer *moment arm*. In physics, the *principle of moments* states that the farther from the *center of rotation* (which for airplanes is the *center of gravity*) that a given force is exerted, the greater its effect at the center. The principle of moments explains why crowbars work so well. This universal physical truth means that mounting an airplane's stabilizers anywhere but at the aft end, which is as far as they can be from the center of gravity, will render them less effective.

Another reason to put them there is *downstream drag*, which enhances stability. Like a weathervane rotating to show wind direction, stabilizers that project into the slipstream tend to align the airplane with the airflow by slightly retarding the rear of the airframe relative to the rest of the machine. Feathers at the back of an arrow provide this same benefit.

It took a while for airplane builders to discover all this. During World War I, for example, the Red Baron's infamous Fokker Dr.I triplane was a bear to fly because it had too short a fuselage and lacked a vertical stabilizer in front of its rudder. The Fokker firm addressed these failings in its next design, the D.VII, which displayed superb handling.

Companies were still learning lessons on this front on the eve of World War II, as Boeing's experience with the Clipper shows. By then, however, it was strictly a matter of aerodynamic fine-tuning; nobody disagreed with the basic ideas of what an airplane's tail should be and do.

Today, triple-tailed beauties such as the Boeing 314 or Lockheed Constellation are a thing of the past. Aerodynamic optimization has pushed the world to single vertical tails.

If all this was so intuitive, why did Wilbur and Orville Wright go such a different route? It all had to do with control in the air.

8 FLIGHT CONTROLS

THE CHARIOT'S REINS

They envisaged the flying machine . . . as a light and living structure, propelled and maneuvered about the sky as if it were a bird possessed.

—SIR CHARLES H. GIBBS-SMITH (1909–82) ON THE WRIGHT BROTHERS[1]

Wilbur Wright arrived in Le Havre in May 1908. Proceeding to the customs house, the forty-one-year-old retrieved a large crate containing a disassembled Wright Model A Flyer built in Dayton and shipped to France the previous year.

With the help of Léon Bollée, an amiable manufacturer of early automobiles, Wilbur took his airplane to Le Mans, some 130 miles (210 kilometers) west of Paris, and arranged to use its Hunaudières racecourse for a flying field. On opening the crate, however, he was distressed to find its contents in disarray, with many parts broken.

Living in Bollée's factory, eating and sleeping beside his machine, Wilbur methodically repaired and assembled this Model A in preparation for the first public demonstration of a Wright airplane anywhere in Europe. He and Orville had left the bicycle business to manufacture airplanes. This was the first phase of their European sales campaign.

The French had heard disquieting rumors for years about the Wright brothers. Apprehension had mounted that these Americans, or perhaps Professor Langley, might have assumed the lead in flight. Consequently, Wilbur's doings were followed with great interest.

Word of the Wrights first emerged in 1902 when Captain Ferdinand Ferber, a French artillery officer who corresponded with Octave Chanute, received from him an illustrated reprint of a presentation that Wilbur had made to the Society of Western Engineers in Chicago in September 1901. The reprint's photos of the Wright 1901 Glider so inspired Ferber that he paid a carpenter to make a copy that, being highly inaccurate, failed to perform.

In April 1903, a major speech by Octave Chanute at France's Aéro-Club de Paris escalated the Europeans' worry to outright alarm. Highlighting the 1902 Glider with compelling photographs, Chanute revealed considerably more about it than the Wrights might have wished. His follow-up article in that November's *Revue Générale des Sciences* put into print astonishing photographs of Wilbur Wright piloting a much larger glider than anyone else had ever flown. One shot even caught him in the act of performing a banked turn.

The French cherished their proprietary claim on human flight. Having given the world the balloon, they felt it was a matter of utmost national pride that they also deliver the airplane. Thus, consternation reigned when conflicting and unverifiable reports arrived that the Wrights had successfully flown a true airplane at the end of 1903. Many U.S. press accounts echoed the spurious claim that the Wrights had flown 3 miles (5 kilometers), a patently ridiculous assertion that caused French experts to dismiss the entire report as unfounded.

Intent on finding out just how much of a lead the Americans had, wealthy Parisian lawyer Ernest Archdeacon commissioned brothers Gabriel and Charles Voisin to build a reproduction of the Wright 1902 Glider. Although proclaimed an exact copy, Archdeacon's Wright-inspired glider, like Ferber's, was far from accurate and scarcely flew. Another French experimenter constructed a glider that came closer to approximating Wright technology, but it too was a failure. Yet a third crashed instantly when towed behind a car. All of this was tremendously reassuring to the French flying community, who took it as proof that the Wright claims were exaggerated.

After their success of December 17, 1903, the Wrights had flown two more seasons in the Dayton area to perfect their technology. They then called a halt to flying and devoted the next two and a half years to building improved engines and seven new airplanes. They also sought to secure customers for their invention. As a result, international press accounts had been reassuringly quiet of late about the Wrights.

In the meantime, imperfect flying machines had begun staggering into the air under power in France, which boasted the largest and most active aviation scene in the world. Alberto Santos-Dumont in his 14-*bis* performed Europe's first officially hailed "flight" in October 1906, although it was actually just a wallowing, marginally controlled hop. In November 1907, a Voisin biplane modified by pioneer aviator Henri Farman set a European record by remaining aloft for more than one minute. Two months later, Farman performed the continent's first circling flight in this same machine.

By the time Wilbur arrived at Le Mans, therefore, the French believed they led the world and largely dismissed the Wright brothers as *bluffeurs*. Wilbur's presence thus became an intriguing puzzle that only deepened as time passed. The unprepossessing American was forthright, modest, unfailingly polite, and radiated quiet confidence. Unfortunately, however, he was also aloof and not much for talking.

Word arrived in July that Glenn Curtiss, a fellow *américain*, had won the *Scientific American* Trophy for the first public display in the United States of an airplane to fly more than a kilometer. The news did not faze Wilbur in the least.

On August 8, 1908, Wilbur Wright was ready at last. A large crowd gathered spontaneously that Saturday. People spread cloths on the grass and settled down to picnics. Others found vantage points on the branches of trees.

A team of helpers wheeled the spindly Wright Model A Flyer out into the sunshine. After some difficulty starting the engine, it roared to life. Looking more like a businessman than an aviator in his gray suit, cap, and high starched collar, Wilbur climbed aboard and gripped the controls.

At his signal, a weight dropped from a derrick, drawing the airplane forward by means of a rope routed through pulleys. Before the astonished onlookers, the Flyer climbed quickly aloft, leveled out at 30 feet (roughly 10 meters), and maneuvered with such authority that it drew gasps. One newspaper noted the next day that the American "took turns with ease at almost terrifying angles and alighted like a bird."[2]

In less than two minutes aloft, Wilbur Wright had stunned an entire nation. Over the following days, he flew longer and more dramatic displays, with figure eights thrown in for good measure. His mastery defied belief. In his hands, the Flyer tracked straight as a locomotive on rails before heeling over into a tight 180-degree turn completed in a few airplane lengths. His machine obeyed his every whim with a degree of precision no European had ever imagined possible.

The French doffed their hats in a generous and ingenuous outpouring of joy. All doubts dispelled, they magnanimously and fervently

A confident Wilbur Wright prepares to fly at Hunaudières racecourse, Le Mans, France, in the summer of 1908.

Horse-drawn carriages wait below as Wilbur Wright flies a passenger at Pau, France, in early 1909.

celebrated the American success. "Wright is a genius," proclaimed Louis Blériot. "He is the master of us all."[3]

"It is a revolution in aeroplane work," agreed pioneer aviator René Gasnier. "Who can now doubt that the Wrights have done all they claimed? . . . We are as children compared to the Wrights."[4]

Ernest Archdeacon's words were perhaps the most telling. "For a long time, for too long a time, the Wright brothers have been accused in Europe of bluff—even perhaps in the land of their birth," said the Parisian lawyer. "They are today hallowed in France, and I feel an intense pleasure in counting myself among the first to make amends for that flagrant injustice."[5]

Six months earlier, thirty-three-year-old Henri Farman had actually managed to land where he took off by skewing his Voisin sideways through the air in a wide, ragged circle performed using rudder alone. Finding and holding a fine balance, he had coaxed the biplane into scuttling sideways like a crab in the air. Each time that the wob-

Henri Farman poses by the tail of his Voisin-built biplane.

bling wings threatened to dip too far, he had backed off on the rudder for fear of capsizing.

That circling flight, Europe's first, had been hailed as a triumph. Now the scales fell from European eyes; the Wrights ruled the air because they alone knew the secret of control.

On October 9, 1890, half a year before Otto Lilienthal began gliding in Germany, France's Clément Ader stood on the grounds of a friend's château. Gravel crunched underfoot as the forty-nine-year-old engineer, balding and moustachioed, readied his startling aircraft for test.

Named *Éole* after Aeolus, ruler of the winds in Greek mythology, Ader's machine combined bat-like wings with a toadstool fuselage that rested on small carriage wheels. A lightweight steam engine drove its forward propeller, the blades of which resembled feathers.

A brilliant self-taught electrical engineer and inventor, Ader made his fortune early in life improving on Alexander Graham Bell's telephone and helping to wire Paris for service. His interest in flight dated back two decades to a gasbag balloon he had built. More recently, George Cayley's airplane concept—as popularized by Henson's Aerial Steam Carriage—had captured his imagination and energies.

Ader's angular *Éole* expressed a clear conviction that emulating natural flight forms was the path to success in the air. With its steam engine now running smoothly, Ader settled into the single seat. Friends stepped back and watched expectantly.

Accelerating at full tilt down the carriage lane, the *Éole* left the ground to perform history's first passage through the air by a manned, self-launched, and self-propelled vehicle. Observers saw the odd craft—which resembled an umbrella more than an airplane—float 160 feet (50 meters) at an altitude of only 8 inches (20 centimeters) before settling back to earth. France's first real aerial steam carriage had hopped but not flown.

It was just as well the *Éole* had not climbed higher into the air. Despite some provision for adjusting its bat wings in flight, this craft offered its pilot no meaningful way to control his course through the air.[6]

British historian Charles Gibbs-Smith, in his day early flight's foremost authority, divided the first generation of aerial pioneers into two distinct camps. Each had a fundamentally different mind-set. Europe's pioneers were *chauffeurs* who envisioned driving around the sky in inherently stable machines that stayed upright in turns. Like Ader, the first chauffeur, they were remarkably cavalier about the whole issue of control. In contrast, the Wrights and those of a similar frame of mind were *airmen*. Like Lilienthal, their guiding light, they sought to learn from the birds and expected to emulate their nimble maneuvers aloft.

As explained in chapter 2, it was all a matter of paradigms. In Europe, Henson's Aerial Steam Carriage led experimenters to assume that the airplane, like the horse-drawn carriage, should stay level as it traveled through the sky and that it would not require active intervention by the pilot unless a change in speed or course was desired.

In contrast, the Wrights' intimate familiarity with bicycle riding predisposed them to thinking it was entirely natural to lean into a turn. Bicycling also led them to expect to participate actively in controlling the airplane every instant that it was aloft.

By 1900, of course, more than just the Henson Aerial Steam Carriage was leading the Europeans astray. It was also the automobile, Europe's newest infatuation. Karl Benz invented it in Germany in 1885. By the time the next century began, the motor car had the freedom of Europe's roads and a burgeoning industry supported its production.

Louis Blériot, a manufacturer of acetylene headlamps, came to aviation from the automobile industry. So did Henri Farman, the Paris-born son of British parents, who raced cars before becoming France's foremost flier and then an airplane manufacturer. Gabriel Voisin, another early flying enthusiast, also loved automobiles; having founded the company that built Europe's first successful heavier-than-air flying machines, he turned exclusively to car manufacture following World War I.

Since cars and airplanes were mechanical vehicles powered by gasoline engines, it seemed natural to equate the two. This mind-set actually predisposed Santos-Dumont, Voisin, and Blériot to think that aviation's challenge consisted primarily of power, not control (still around, Ader himself claimed a lack of power was the only reason he had not done better). Just get an airplane off the ground, this prevailing attitude said, and the rest will take care of itself.

This was certainly the view of Léon Levavasseur, the talented artist-engineer who gave the world engines and airplanes named Antoinette. The former combined light weight with more power than the Wright brothers had or needed. As for the latter, Hubert Latham's exquisite mount was the world's first successful monoplane and one of the most beautiful flying machines of all time.

In the United States, Samuel Langley shared this conviction so fully that he devoted most of his research funds to developing a powerful engine. Fortunately, the Wrights had a different take on flight. From the outset they were preoccupied with controllability. In their minds was the death of Otto Lilienthal.

Germany's flying man had controlled his gliders by flailing his legs and upper body in the direction he wanted to go. The Wrights knew that in scientific terms what Lilienthal was doing was shifting the position of the glider's center of gravity relative to its *center of lift*. These were the respective points where aggregate physical forces—gravity and aerodynamic lift—acted on his glider, the former as a lever and the latter as a fulcrum.

When the center of gravity overlapped with the center of lift, Lilienthal's glider was in balance and flew straight. But when Germany's airman threw his legs to the right, the center of gravity shifted in that direction and the wing dipped into a turn. With lift's upward push now to the left of the center of mass, the vehicle naturally fell off to that side.

To arrest this turn, all Lilienthal had to do was shift his weight the other way. And to lower the nose, he had but to throw his body and legs forward, placing the center of gravity ahead of the center of lift. It was a simple means of control that anyone could intuitively understand.

The Wrights understood it, but they also saw the meaning of Lilienthal's death. That tragedy's stark lesson was that this system of control did not suffice even for a small single-person glider. If flight was ever to be practical, a better method was needed, one that could be scaled up along with the machine itself. As Wilbur wrote in the sum-

Otto Lilienthal controlled his gliders by shifting his weight to change the craft's center of gravity.

mer of 1899, the "problem of equilibrium constituted the problem of flight itself."[7]

But what should this alternative mechanism be? English engineer Francis Wenham, whose work the Wrights knew and admired, provided a clue. Where others looked to a ship-style rudder for control, Wenham had said in 1866 that turns in flight should be accomplished by generating more lift on one of an aircraft's wings than on the other.

Wenham based this insight on his observation of birds. Although the exact mechanism was unclear, he saw that they controlled their direction of flight primarily with their wings. Rudders were fine for turning ships, he believed, but to turn an aircraft one should look to the wings. It was prescient advice. So closely would Wilbur and Orville follow it that their initial gliders had no tails at all.

Ironically, just as these Ohio brothers took up flight's challenges in 1899, another dreamer died as a result of his gliding experiments. Working in Scotland, British pioneer Percy Pilcher had built several different designs and was on the verge of testing a powered hang glider when he succumbed to injuries sustained in a crash. He was thirty-three years old.

The world was awash with ideas about flight. These came from all quarters and struck harmonic chords on emotional as well as cerebral levels.

Take Louis Mouillard, a Frenchman living in Algeria who studied birds and made a few gliding experiments in the 1860s and 1870s. In his influential 1881 book *L'Empire de l'Air*, Mouillard advocated gliding as the path to learning. When Wilbur wrote to the Smithsonian Institution in May 1899 requesting information about flight, he received an English translation of this work that the Smithsonian had published in 1893.

Writing passionately and poetically, Mouillard—whose own gliding experiments were insignificant—spoke to the soul as much as the mind. His words fueled the Wrights' growing desire to build and fly gliders even if all it amounted to was a bit of fun. But the Frenchman himself predicted it would be more than that.

"If there be a domineering, tyrant thought," Mouillard wrote, "it is the conception that the problem of flight may be solved by man. When once this idea has invaded the brain, it possesses it exclusively. It is then a haunting thought, a waking nightmare, impossible to cast off."[8]

To understand what the Wrights and their contemporaries grappled with, it is helpful to take a brief look at the airplane as we know it today. All airplanes have three axes of motion: *pitch*, *roll*, and *yaw.* Mutually perpendicular, these axes intersect at the airplane's center of gravity, which is a point within the fuselage between the forward part of the wings. The center of gravity can shift fore or aft depending on how the airplane's load is distributed. As long as it remains within defined limits, however, the airplane will fly properly.

The pitch axis can be pictured as a side-to-side horizontal line running more or less in line with the wings. When an airplane rotates about this axis, its nose tilts up or down. In contrast, the roll axis runs fore and aft through the fuselage. When an airplane rotates around this axis, its wings dip to the left or right. As for the yaw axis, it runs vertically. When the airplane rotates around this axis, its nose swings sideways left or right.

All airplanes have control surfaces that deflect in the slipstream to alter the course of flight. The elevators are horizontal surfaces at the rear that control pitch. Mounted outboard on the trailing edges of the wings are the ailerons, which control roll. The rudder, which controls yaw, is the tail's vertical control surface.

On a typical airplane, the control wheel (or stick in the earlier days) controls pitch and roll. Fore and aft inputs make the airplane descend or climb, while left or right inputs cause it to bank. As for yaw, the rudder pedals control that.

As all pilots know, trying to turn a typical light airplane using just the rudder pedals (yaw only) causes the airplane to respond very sloppily. After a lag, the wing will drop because of a coupled rolling moment. Thereafter, constant rudder corrections are needed to hold the wing at the desired angle of bank.

Trying to turn using just the wheel (roll only) also results in sloppy behavior. The nose initially slides to the wrong side before coming around and banking in the desired direction. This phenomenon is called *adverse yaw.*

The secret to smooth, coordinated turns in the air is using both wheel and rudder at the same time. This banks the plane smoothly without adverse yaw. It also keeps the plane flying directly into the slipstream instead of inefficiently skidding or slipping.

Lastly, all pilots understand the distinction between stability and control. That they are separate subjects, albeit closely related, is today self-evident.

All of this sounds perfectly reasonable and logical today. At the turn of the last century, however, it was a different story. Nobody knew enough even to be able to agree on the definitions of key concepts, let alone formalize the requirements for achieving control in the air.

Of all the challenges inherent in inventing the airplane, controllability was the most baffling. Solving it was pure detective work.

George Cayley thought he understood controllability, but his designs made provision for pitch and yaw control only. Being analogous to a whale's flukes and a ship's rudder, those flight controls were the obvious ones. Flying his model gliders delighted the Yorkshire baronet. At times, however, they probably also brought a puzzled frown to his face. This would have happened when he launched a glider with its cruciform tail canted left or right. Instead of turning like a ship, the model would have swung to the desired side, fallen off in a bank, and crashed. Why did this happen? What did it augur for a future form of transport that Cayley alone imagined at the time of the Napoleonic wars?

Based on Cayley's thinking, the Henson Aerial Steam Carriage also featured control around two axes. William Henson's machine—designed in detail as a full-size airplane but built only as a model—lacked any mechanism in the wings for roll control.

Otto Lilienthal likewise had two-axis control, although his two were pitch and roll (with fixed vertical tails, his gliders had no provision for yaw control). But pitch and yaw were the most common combination. Alberto Santos-Dumont's 14-*bis* of 1906 and Voisin's biplanes of the following years featured it.

French pioneer Robert Esnault-Pelterie developed an advanced but only marginally successful monoplane in 1907. This machine featured roll control (via downward-only wing warping) and yaw control but no pitch control. A fascinating machine, the R.E.P. (taking its name from his initials) flirted with steel-tube construction, cantilever wings, and hydraulic brakes before World War I.

Ultimately, Europe's experimenters were groping in the dark when it came to flight control.

Wilbur Wright read a book about ornithology and another addressing the mechanics of birds' flight. He visited rocky outcroppings overlooking a river and studied soaring hawks by the hour, but they were too distant to betray useful insights. As for small birds in the Wrights' Dayton neighborhood, they flew too quickly for study.

Beginning an intellectual journey, the brothers speculated that birds control their direction of flight by shifting their weight and drawing in a wing to temporarily reduce lift on one side, thus tilting them to that side. Then one day in West Dayton, Wilbur saw a pigeon perform such a display of aerial artistry that these theories went out the window.

Wilbur realized that birds turn by twisting their wingtips to temporarily generate more lift on one side than the other. No shifting of weight, no readjustment of relative wing area, just a dynamic change in each wing's pitch or *angle of incidence* that affected how big a bite of the air it took from one moment to the next.

Here, then, was how an airplane also should turn. One wing should temporarily create more lift and the other less to tilt the machine into a bank that would end with an opposite application of control forces. But how to achieve this?

The Wrights toyed with the idea of building a glider employing shafts meeting at central gears to mechanically pivot the wings to opposite orientations. Seeing no way to keep the weight sufficiently low, however, they abandoned the concept.

Then one day Orville returned home to find Wilbur brimming with triumphant excitement. He held forth a small, empty cardboard box that had come off the bicycle shop's supply shelf from which he had removed its two narrow ends. Imagining two parallel long sides also gone, it roughly approximated the wings of a Chanute-style biplane glider.

Holding the corners of each end between finger and thumb, Wilbur gave this box a helical twist. The parallel "wings" distorted, angling up on one side and down on the other. Here was the solution to the control problem.

Wilbur's idea was *wing warping*. By leaving intact a Chanute biplane glider's lateral trussing but removing its fore-and-aft trussing and then installing adjustable tensioning cables for the latter, a glider could be built whose wings could be dynamically adjusted in flight. On one side, they would take on a greater angle of attack; on the other, they would flatten out. With more lift on one side, the glider would turn.

On a windy day in July 1899, Wilbur—accompanied by two small boys—went to a field in Dayton to test a biplane kite built to this formula. It had superimposed wings spanning 5 feet (1.5 meters) and a trailing fixed surface. To control it, Wilbur held two sticks, each secured at both ends with long cords.

Holding the kite aloft, the boys released it at Wilbur's signal. It rose into the wind, and Wilbur tried out his wing warping. The kite obediently tilted one way or the other on command. Mechanically achieved roll control had been demonstrated.

The demands of the highly seasonal bicycle business preempted further work until August 1900. Wilbur, who at that point was more interested in flight than his brother, worked up the design and set to work constructing a man-carrying glider. Unable to find spruce in Dayton's lumber yards sufficiently long to serve as wing spars, he built what he could. The spars he would obtain in North Carolina.

From replies to his query letters, Wilbur had settled on Kitty Hawk as offering the strong and constant winds, flat open ground with available hills, and cushioning sand he had defined as requirements for his gliding experiments. Gliding into a strong wind would allow testing at low ground speeds since the glider's forward progress through the air would be subtracted from the wind's velocity. But that would work only if the wind was relatively constant and free of eddying currents, and that in turn required a sea breeze.

Arriving there in September, Wilbur found wood for spars, but it was pine, not spruce, and shorter than he had planned. It would have to do. Working in Kitty Hawk, he took the ribs, struts, wires, and fittings he had already prepared and completed the glider, covering it with a fine-weave fabric. Orville arrived at the end of the month and the brothers set up a camp at Kill Devil Hills, a remote area four miles south.

The "hills" were three bare sand dunes, one higher than the rest. This biggest was called Kill Devil Hill, leading to the confusing situation of both a singular and plural of the same place name being in use. Beyond the hills a flat expanse of open sand extended left and right for a considerable distance, bordered by the Atlantic Ocean to the northeast.

The Wright 1900 Glider had a wingspan of 17 feet (5.2 meters) and featured a forward elevator. Wilbur may have placed it there because a shorter and simpler control linkage was required than for a rear elevator. Since they were there to learn, he may also have opted for a forward location because he wanted to observe the elevator's performance in flight. Whatever the reason, this glider set the pattern for all subsequent Wright machines and also influenced European practice.

Tested throughout October 1900, this machine was flown primarily as a kite controlled from the ground because it generated so much less lift than hoped. Nevertheless, before striking camp at the end of the month, the brothers dragged it over to Kill Devil Hill, where Wilbur made a dozen or so free flights, providing at least a brief taste of flying.

Although the 1900 Glider flew poorly and was less than controllable in gusty conditions, its elevator worked well for pitch control. Wilbur found he could skim low, staying just above the sand and controlling his height precisely to set gently down where desired. As for the wing warping system, it worked well in tests from the ground, but they disabled it for Wilbur's glides. Tests of this lateral-control system would have to wait for the next season's flying with an improved model.

Kitty Hawk's residents next saw the Wrights in the summer of 1901, when they returned for a longer flying season. They assembled their second manned glider right at camp. This new machine was similar overall but larger than the previous glider. It had a wingspan of 22 feet (6.7 meters), a chord of 7 feet (2.1 meters), and 75 percent more lifting area. With a total weight of 98 pounds (44.5 kilograms), it could be carried by two men, although trudging up a sandy hill with it was hot, tiring work.

Like the previous machine, the Wright 1901 Glider had no tail and the center of its lower wing had a gap so that Wilbur could stand up until ready to fly. At that point, with Orville and a local helper running alongside to steady the wingtips, Wilbur would tuck his legs up and hook his ankles over a supporting spar.

The Wrights were disappointed to find the new glider's lift and longitudinal stability were not as good as those of the previous year's machine. Performing modifications at camp, they reduced the wing camber and stiffened its wings overall, which they realized were distorting under the pressures of flight.

Thus revised, the glider flew quite well. Many flights exceeded 300 feet (91 meters) in length, and the longest stretched 389 feet (119 meters). Despite these achievements, however, their troubles were far from over, because a bizarre new problem appeared when they finally tested wing warping in flight.

On the Wright 1901 Glider, wing warping was controlled by the feet, which rested on a T-bar as the pilot lay prone on the glider's wing. It was an ingenious system. To bank to the right, for example, the pilot would press the pivoting bar with his right foot, causing wires routed through pulleys to impart a temporary helical twist to the wings. Under the tension of this control input, the wings on the glider's left side would angle upward while those on the right side flattened out. The resulting difference in angles of attack, and thus aerodynamic lift, would roll the machine into a right-hand turn.

At least that's what was *supposed* to happen, and it had worked in kite tests. Now, however, incomprehensible things sometimes occurred when Wilbur attempted to bring up a low wing or turn one way or the other. The glider would tilt as commanded but then rotate alarmingly the other way. Slipping sideways out of control, it would crash to the sand, resulting in splintered wood, consternation, and the need for repairs.

Talking it through in camp, the brothers correctly diagnosed the problem. During warping, the side of the glider generating greater lift was also experiencing greater drag even as the opposite side experienced reduced lift and drag. Moreover, the wings on the outside of a turn must travel farther and thus fly faster than the inboard wings. Since the amount of lift and drag is proportional to the speed with which a wing passes through the air, this disparity was accentuated in turns.

The net result was forces working in opposition. Even as lift rolled the glider into a turn around the longitudinal axis, drag was wrenching it out of the turn by yawing it the other way around the glider's vertical axis. Trying to bring the low wings up using warping also risked commanding them to too high an angle of attack for their reduced airspeed, causing them to stall and lose their remaining lift.

That explained the odd behavior and crashes, which scuttled the brothers' hopes for an easy solution to the controllability problem. The flying season again having come to a close, the Wrights left Kitty Hawk thoroughly dispirited. In addition to the vexing control issues, their gilder was generating far less lift overall than it should have. On the long train trip home, the brothers discussed abandoning their research, although they knew they could not. The challenge was simply too fascinating.

Instead, they spent the closing months of 1901 learning from their wind tunnel.

In September 1902, the Wrights returned to Kitty Hawk armed with new knowledge and many crates containing the parts of a large new glider. Assembled in eleven days, the Wright 1902 Glider was longer, narrower, and had wings of a flatter camber. Wingspan was 32 feet (9.75 meters), chord 5 feet (1.52 meters), and weight 112 pounds (51 kilograms).

The 1902 machine featured the same wing warping mechanism as the previous year's glider. However, this system was no longer actuated by the feet. Instead, the pilot would bank by shifting his hips to the desired side in a wooden cradle.

For the first time, there was a tail. Mounted on struts aft of the wings were two fixed vertical vanes that the brothers hoped would cure the previous year's control issues. By resisting side-to-side swings, they believed, this vertical stabilizer would help the glider track through turns without experiencing as great a disparity in airspeed from one side to the other.

From the outset, the 1902 machine flew long distances. Demonstrating lower drag and higher lift, it closely matched the brothers' computed expectations. However, they also found that wind gusts from the side lifted the glider's upwind wings, upsetting its course through the air.

The brothers addressed this sensitivity by revising the glider's diagonal trussing so as to give its wings a slight downward arch as viewed from the front. Accomplished at their camp, this rigging change made the glider's wingtips droop 4 inches (10 cm) relative to the center of the wings.

Thus revised with anhedral, the glider became relatively insensitive to side gusts. Anhedral also eliminated what little lateral stability there had been, but the Wrights felt this trade a positive one. Lacking fixed horizontal stabilizers front or back, this machine—like all the Wright gliders and powered flyers—was also not stable fore and aft.

The lack of lateral and longitudinal stability meant that Wright pilots would have to be on their toes, actively controlling their mounts every instant while aloft. But this did not bother them for an instant. After all, bicycles had to be controlled all the time or their riders were in for a spill, yet nobody had trouble riding them. And with flights generally so brief (in those early days, at least), there was no time for pilots to become fatigued, so what did it matter?

Octave Chanute never understood or trusted the Wrights' approach. He felt airplanes should be inherently stable for overall safety. So did the Europeans, who from the outset sought to build intrinsically stable flying machines. But stability fights against controllability, and for the Wrights controllability was the secret to flying.

Wing warping now worked and all seemed well. Gliding into the wind, Wilbur even demonstrated an S-turn, bringing the 1902 Glider broadside to the wind first one way and then the other, an unheard-of feat for a glider. But as testing continued, control instability again cropped up, and this time it was worse. In fact, in Orville's own words, it was "absolutely dangerous."[9]

As before, the positively warped wings would suddenly drop with no warning, and the glider, slewing violently, would fall sideways "as a sledge slides downhill or a ball rolls down an inclined plane, the speed increasing in an accelerated ratio," as Wilbur described it.[10]

They had figured out what was going on with wing warping the previous year. Now they saw that the vertical stabilizer was aggravating the issue rather than helping. In Orville's words, the tail "caused one wing to be checked and the other to be speeded up."[11]

By resisting sideways motion, this fixed tail was aerodynamically preventing the glider from yawing and instead whipped it into a violent sideslip. Out of control, the glider would spin down around its low wing, invariably gouging a circular trough with its wingtip.

This signature in the sand led the Wrights—who just weeks before had dug a well for their camp—to name this sobering phenomenon *well digging*. There were no injuries from these crashes, but they kept the brothers busy with repairs.

On the evening of October 2, Wilbur went to bed early, as was his habit. Orville typically puttered around and read before finally turning out the lantern. This night he lay on his cot, listening to the mingled melody of crickets and surf. Sleep defeated him because he had drunk too much coffee.

As he lay awake, the younger Wright kept replaying in his mind their glider's odd behavior. He traced mentally the different forces coming into play. Suddenly the solution came in an intuitive flash: replace the fixed vertical stabilizer with a movable rudder. During a turn, this rudder would deflect to the side of the low wing, pivoting the glider so that it tracked properly through the turn. This control input from the rear would also push the high wings around, overcoming the higher drag on that side of the glider.

The brothers made this change, wiring the rudder into the warping mechanism so that it too was actuated by the hip cradle. From now on, the rudder would automatically deflect when wing warping was commanded.

The Wright 1902 Glider made between seven hundred and eight hundred glides following modification with a movable rudder. Not once did Wilbur or Orville encounter further difficulties in control. In all, they would log nearly a thousand flights at Kill Devil Hill during September and October, making it by far the most productive of their three stays to date in North Carolina.

Here was history's first fully controllable aircraft (lacking power, it was not an airplane). Never before had any man-made device flown with control around all three axes.

Flights of 500 feet (150 meters) were common, and some exceeded 600 feet (180 meters). The longest was 622 feet (189.6 meters), and the greatest duration aloft was twenty-six seconds. In all regards, the 1902 machine matched its predicted performance, attesting to the tabular data amassed by the Wrights and giving them confidence that they could predict the performance of future machines.

Based on this crowning success, the Wrights filed a patent in March 1903 describing their control system as worked out on the 1902 Glider. It speaks volumes about the emphasis they put on control that their patent was based on a glider rather than the powered machine that followed.

In 1904 and 1905, Wilbur and Orville arranged to use an almost 100-acre (40-hectare) meadow at Simms Station eight miles east of Dayton to continue their experiments. Called Huffman Prairie, the property was bordered on two sides by well-traveled roads. Beside the Dayton-Springfield Pike, the busier of the two, ran an interurban light-rail system.

While the Wrights valued privacy so as to be able to work, they were not secretive. In fact, one of the first things they did at Huffman Prairie was to invite local reporters to witness their initial flights. Alas, lack of wind and a very balky engine kept them from getting into the air. The event was restaged soon afterward, but the same thing happened, and the reporters never came back.

Had these attempted press demonstrations succeeded, the world would have had electrifying news of the airplane's certain invention in the middle of 1904. Instead, locals would watch the Wrights fly day after day, taking it entirely for granted, even as the world at large dismissed as unfounded rumor the first great invention of the twentieth century.

The Wrights traveled to Huffman Prairie and back via the light-rail system. It was a forty-minute trolley ride each way from their home in West Dayton to the Simms Station stop, diagonally across from their flying field. At Huffman Prairie, the Wrights built a shed like the ones they had constructed at Kill Devil Hill. Here in May 1904 they completed an airplane that was similar overall to the Wright 1903 Flyer.

Heavier and more robust, the Flyer II also had a new engine providing one-third more power. Ready to fly, they found themselves waiting instead. Most days, Dayton's variable weather and capricious winds—particularly the lack of it—kept them grounded in frustration.

In early September, they improvised a derrick with a hanging weight that could be dropped to catapult the airplane aloft. This eliminated the need for a brisk wind into which to take off. After that, flying became frequent.

The brothers found that the 1904 machine shared its predecessor's pronounced dynamic fore-and-aft instability. Wanting to porpoise in flight, it required a skilled hand on the elevator to keep it flying straight. Bobbing oscillations arose to plague many flights, resulting in hard landings punctuated by splintered wood.

The season's great accomplishment came on September 20, when

Wilbur executed the world's first-ever circling flight. This was a necessary capability if they were to stay within the confines of their small pasture, which they intended to do for safety. Caution also led them to keep the airplane flying low over the ground, generally no higher than the top of their shed.

Amos Root, an elderly and energetic apiarist, had been invited to witness that event. On January 1, 1905, he printed the first eyewitness account ever published of an airplane flight in, of all places, his journal *Gleanings in Bee Culture*. "Dear friends," Root began, "I have a wonderful story to tell you—a story that, in some respects, outrivals the *Arabian Nights* fables."[12]

The miraculous machine Root saw put through its paces left him awestruck. "When it first turned that circle and came near the starting-point," he told his readers, "I was right in front of it; and I said then, and I still believe, it was one of the grandest sights, if not the grandest sight, of my life."[13]

At the end of the 1904 season, the Wrights knew they had more work to do; their control of the flyer was incomplete. In a turn, it sometimes kept dropping a wing despite every effort to the contrary. As Orville succinctly put it, "the problem of equilibrium had not as yet been entirely solved."[14]

On July 18, 1905, at Billancourt, France, in Paris' western suburbs, a glider on pontoons bobbed in the Seine River. Designed by Gabriel Voisin with ideas contributed by Louis Blériot, who had commissioned Voisin to build it, this craft was a Hargrave box kite with short biplane wings of unequal span.

Between the wings on each side were two vertical fore-and-aft fabric panels. The outboard "side curtains" were canted to connect the tips of the shorter-span lower wing to those of the longer-span upper one. At the rear, a biplane tail unit was likewise enclosed top, bottom, and sides like a Hargrave kite. Voisin counted on these design features to lend his glider inherent lateral, longitudinal, and directional stability.

The Voisin-Blériot glider also showed the influence of the Wright brothers, whose galvanizing success with the 1902 Glider had reawakened Europe's determination to be first in flight. Most noticeable was this glider's front elevator, which all European biplanes would sport for some years because of the Wrights.

Gabriel Voisin climbed into the glider's seat. At his signal to the crew of an idling motorboat, its engine roared and a tow rope came taut. The glider accelerated smoothly across the water, lifted free, and immediately dipped a wing. Falling into an uncommanded sideslip, it broke up on impact.

Voisin came close to drowning. The reason was one-axis control: aside from its front elevator, the Voisin-Blériot float glider of 1905 incorporated no mechanism for guiding its course through the air.

Between late June and mid-October 1905, the Wright brothers logged more than forty flights at Huffman Prairie in a new airplane fitted with the previous one's engine. The Flyer III, also called the Wright 1905 Flyer, differed noticeably from its predecessors by having its front elevator farther forward and its rear rudder farther aft.

Flown at Huffman Prairie in 1905, the Wright Flyer III was history's first practical airplane. Its longest flight came on October 5, when Wilbur covered nearly 25 miles (40 km) in thirty-eight minutes aloft.

This fore-and-aft lengthening was made after Orville lost control of the craft on July 14. The machine crashed nose first, its elevator and supporting outriggers absorbing some of the impact. Bouncing down the field, it slid upended to a jarring stop that shot Orville through the shattered top wing. He was battered and dazed but, fortunately, unhurt.

The brothers rebuilt the craft with an enlarged elevator set farther forward. This change significantly improved the flyer's longitudinal stability and handling. Eliminating the problem of occasional losses of control, it also cleared the way for remarkable successes as the flying season wore on.

One problem still remained, however. The Flyer III continued to exhibit an alarming tendency for the wings on the low side to continue to drop during banks. On one September flight, the machine had turned uncommanded into the tree it was circling. Torn branches adorned its struts when it landed.

What was happening? The low wings, being on the inside of a turn, were slowing down and generating reduced lift. When the pilot applied opposite warping to bring up these low wings, the combination of a high angle of attack and too little forward airspeed caused them to stall out, precipitating a further drop.

Given sufficient altitude, control was regained by putting the nose down for more flying speed and then bringing the wings level with wing warping, which remained effective throughout on the high side. However, the Wrights lacked sufficient altitude for this to be an option because they were staying low for safety.

The solution turned out to be to unlink the rudder from the wing warping, thus allowing its independent use. In banks when the low wing would not come up, the application of "opposite rudder" or "top rudder"—meaning a deflection of that control surface to the high side—brought up the low wing.

With this last issue resolved, handling was excellent. Tight turns and figure eights were flown with impunity. On September 26, with his father watching, Wilbur made circle after circle, covering 11 miles (18 kilometers) in twenty minutes, landing only because he ran the gas tank dry. On October 4, Orville flew more than 20 miles, and the next day his brother flew just shy of 25 miles (40 kilometers) in thirty-eight minutes.

Here at last was a manned, powered, heavier-than-air vehicle that was sturdy, reliable, maneuverable, and could remain aloft for extended periods. If the Wright 1903 Flyer was the world's first true airplane, the 1905 Flyer III was the first *practical* airplane. It thus ranks as of nearly equal historical significance. This priceless artifact is today displayed at Carillon Historical Park in Dayton, Ohio.

In January 1906, the respected French journal *L'Aérophile* published the basic text with illustrations of the Wright brothers' pending U.S. patent. Rejected as originally submitted by Wilbur and Orville in March 1903, the application was revised with the help of a patent attorney and resubmitted in 1904. The U.S. government would grant this patent in May 1906.

The article included a description, in the Wrights' own words and bolstered by helpful diagrams, of their three-axis control system. It explained the concept of lateral control based in the wings (some-

Afraid of falling off to either side, Gabriel Voisin equipped his biplanes with fabric panels between the wings to prevent sideslips.

thing no Europeans yet considered) and described how this control, in coordination with use of the rudder, brings about smooth turns in the air.

Here, a full two and a half years before Wilbur's flights at Le Mans and two years before Farman's wobbling circuit, was the Wrights' secret laid bare for all to see. Although *L'Aérophile* was required reading for Europe's active early flight community whose epicenter was Paris, not one experimenter saw the obvious.

Why not? Here again we see the tyranny of a reigning paradigm. Instead of a mechanism for roll control, the Europeans sought sources of inherent stability that would keep airplanes from tilting laterally. This was why Alberto Santos-Dumont gave his 14-*bis* of October 1906 so much dihedral.[15] It was also why Gabriel Voisin and others often equipped their biplanes with sideslip-inhibiting vertical fabric panels aligned fore and aft between the wings.

Yes, Europe's chauffeur mind-set had effectively blinded one and all to the truth. Farman and Léon Delagrange, France's two top aviators, specialized in flying cross-country because setting distance records did not require turning. Others too accepted and operated within the constraints of inadequate control. Then Wilbur arrived to unchain their minds.

July 4, 1908, dawned breezy with heavy overcast in Hammondsport, a small town in upstate New York near the Canadian border. It had rained much of the day, dampening Independence Day spirits, but by late afternoon it started to clear. Now as dusk approached, raking sunshine turned the surrounding hills a vivid green against black clouds.

On a long grassy slope near town, Glenn Hammond Curtiss checked his *June Bug*, a spindly biplane with yellow-tinged wings that bowed together. Barely thirty years old, Curtiss had personally designed this airplane and its engine, which he built with the help of the staff of his engine company. Slight of stature with a perpetual frown and piercing blue eyes, Glenn Curtiss was Hammondsport's most famous resident, a former daredevil motorcycle racer turned businessman, engine maker, and now pioneer aviator.

Assorted friends, helpers, and official observers clustered with him

Pioneer aviator Glenn Curtiss and friend.

around his airplane. Dominating this inner circle was Alexander Graham Bell, Curtiss' elderly patron, whose ebullience was audible at a distance. The town's citizenry and others watched from cloths thrown on the grass, which sloped gently away for more than a mile. Flags waved, a small band played, and children darted here and there laughing and shouting. A thunderstorm pushed through, briefly drenching the scene, but nobody seemed to mind.

The reason for this unusual holiday activity was a $2,500 prize and solid silver trophy offered by *Scientific American*, a leading publication, for the first person to complete a public, officially witnessed airplane flight of more than 1 kilometer in the United States. Curtiss had seen pictures of the trophy, which depicted an eagle atop a globe with winged horses ringing its base. He meant to claim it and had chosen this time and place.

The magazine had posted this prize in part to make amends to the Wright brothers, whom its staff had too long ignored. To *Scientific American*'s consternation, however, the Wrights—aloof and dismissive of anything smacking of showmanship—passed up this golden opportunity despite having made unofficial flights before invited guests and passersby of up to 40 kilometers back in 1905.

Curtiss flies his June Bug *to win the* Scientific American *trophy, July 4, 1908.*

Right now Wilbur was in France preparing to fly. As for Orville, he was readying a different Model A for demonstration to the U.S. Army near Washington, D.C., in September. This left the field wide open for Glenn Curtiss. Conditions being favorable, Curtiss climbed into the *June Bug*'s seat and set his shoulders in the yoke that controlled triangular ailerons at the airplane's wingtips.

The propeller was spun and the engine caught with a bellow. Breaking free of the ground, the *June Bug* wallowed unsteadily through the air in a straight line. Curtiss passed the 1-kilometer flag marker and kept going. Only when hemmed in by looming trees at the end of the meadow did he set down, unwilling or unable to attempt a turn. He had flown just over a mile (1.6 kilometers) to claim America's first aeronautical trophy.

The *June Bug* was a poor airplane by all accounts. Curtiss demonstrated great bravery just flying it. Although it nominally had the Wrights' idea of three-axes control, Curtiss put little faith in the plane's ability to do more than stay on an even keel. The *Golden Flyer*, his next design, was far better. That classic pusher biplane—prototype of the famous Curtiss Model D series—featured ailerons mounted between the wings. Curtiss mastered control in the air in this later type, setting the stage for his win at Reims in 1909.

At the start of 1909, Wilbur Wright relocated to Pau in the south of France to train French pilots. Shortly after arriving there, he was joined by his sister, Katharine, and a convalescing Orville. On September 17, 1908, while demonstrating the Wright Military Flyer to the U.S. Army at Fort Myer, Virginia, the younger brother had crashed when a propeller split, breaking the airplane. His passenger—Lieutenant Thomas Selfridge, a colleague of Glenn Curtiss' in Alexander Graham Bell's Aerial Experiment Association—was killed, gaining the unfortunate distinction of being the world's first aviation fatality.

At the start of April, the three Wright siblings relocated to Rome, where the brothers showed off their airplane to Italy. They remained a month before returning home in triumph. In July of that year, Orville returned to Fort Myer to complete his demonstration to the U.S. Army of what became history's first military airplane. Thereafter, Orville traveled to Germany for flight demonstrations at Berlin and Potsdam through the end of October. Today displayed by the Deutsches Museum, the German-built machine he used there is the only surviving Wright Model A.

The Europeans and British learned very quickly from the Wrights. Almost immediately, the continent's airplanes began flying with roll control added to their wings for full three-axis control. This was achieved with either wing warping (changing the camber of the wing itself) or ailerons (a movable surface attached to the wings).

Competing to be first across the English Channel in 1909, Hubert Latham flew an Antoinette equipped with ailerons, whereas Louis Blériot used wing warping on his Model XI. Alberto Santos-Dumont's graceful Demoiselle employed wing warping, whereas Farman used ailerons (the latter technology would soon win out because the former works only on thin, flexible wings and requires drag-inducing wires).

The only European who did *not* learn from the Wrights was Gabriel Voisin. He willfully resisted out of Gallic pique. Consequently, of the many machines flying at the Reims air meet in August 1909, his alone lacked any provision for roll control.

If the Wrights had lessons to teach others about roll control, it was the other way around when it came to pitch. Orville belatedly acknowledged as much in a letter he wrote to Wilbur in September 1909. "The difficulty in handling our machine is due to the [horizontal] rudder [i.e., elevator] being in front, which makes it hard to keep on a level course," he wrote. "I do not think it is necessary to lengthen the machine, but to simply put the [horizontal] rudder behind instead of before."[16]

The result was the Wright Model B of 1910, a revised Model A with its elevator at the rear. America's first production airplane, the Model B was also the first flyer with wheels. These were merely affixed to the skids, however, and not a new landing gear. The Model B retained the Wrights' front blinkers, but they were now mounted on the front skid struts since there was no forward elevator. By then, however, the rest of the world had pretty much agreed that the vertical stabilizer should be at the rear with the rudder.

In short, the world had passed the Wrights by. For whatever reason—be it an excessive investment in the past or too much energy expended on their patent disputes with Glenn Curtiss and the Europeans—they were now far behind on the configuration front.

Europe also had lessons to teach the Wrights about stability. The winning formula for aviation actually turned out to be a combination of the Wrights' three-axis control concept and Europe's pursuit of inherent stability. This synthesis made possible airplanes that were reasonably safe, stable, and forgiving, as well as fully controllable.

Only in the 1970s would aviation return to the Wrights' notion of inherent instability as the route to greater maneuverability. Starting with jet fighters such as the General Dynamics (today Lockheed Martin) F-16 Fighting Falcon, modern fighters offer the best of both worlds. Designed statically unstable for extreme maneuverability, they nevertheless present their pilots with stable, forgiving handling characteristics thanks to computerized fly-by-wire flight control systems that do the hard work. So inherently unstable are these astonishing machines that they could not be flown without the aid of computers.

Jetliners too have a noteworthy controllability feature not seen in the early days of aviation. These are the *spoilers* atop the wings. Hinged at front, they can rise fractionally into the slipstream in flight. Depending on how they are employed, spoilers can do several things. When used differentially so that only the spoilers on one side rise into the slipstream, they tilt the airplane into a bank. Like ailerons, therefore, they provide roll control (spoiler use is in fact preferable in some instances because it avoids the adverse yaw associated with ailerons).

When used symmetrically, spoilers either slow down the jetliner without changing its altitude and attitude, or allow it to descend at a high rate without excessive speed buildup. It all depends on how the flight crew uses the spoiler application. They can also be actuated by the airplane's autoflight system when on autopilot.

Because of their placement on the wings, a jetliner's spoilers are readily visible to the passengers and can be fun to watch. They scarcely rise when used for roll control but are more evident when employed for killing lift. The laminar flow over the wing becomes locally turbulent when lift is spoiled. Inside the cabin, passengers perceive this as a slight shuddering somewhat reminiscent of driving fast over a rough road. Because spoilers sacrifice momentum or altitude created by consuming fuel, they are used sparingly. However, all jetliners are aerodynamically so extremely clean that they need this capability.

The most dramatic use of spoilers is on landing, when they pop up more fully than they can in flight to help slow the airplane. This application of ground spoilers is one of three separate and independent braking systems that jetliners have for decelerating, the others being wheel brakes and thrust reversers. Ground spoilers contribute to stopping in two ways: they provide direct aerodynamic braking, and they kill off the remaining wing lift to drop the airplane's weight onto its wheels so that its wheel braking is more effective.

For safety, ground spoilers are inhibited in flight. Until signals reach the jet's computers telling it the airplane has landed, the ground spoilers are not allowed to deploy.

9 FLIGHT DECK

COCKPITS FOR AERIAL SHIPS

Deep into the darkness peering; long I stood there wondering, fearing . . .

—"THE RAVEN" BY EDGAR ALLAN POE (1809–1849)

As human beings built their flying machines in the opening years of the last century, they gave them open *cockpits* reminiscent of the wells from which a small sailboat is controlled (indeed, this is where aviation drew the term from). Instead of sheets, a tiller, and a compass, however, these aerial vehicles surrounded their operators with flight controls and instruments.

Early on, people had different ideas of how human beings should physically occupy their airplanes. The Wrights flew lying down, oriented like birds in flight to reduce air resistance. This changed in 1908 with their welcome adoption of upright seating.

When Brazilian aviation pioneer Alberto Santos-Dumont performed Europe's first officially witnessed hop by a powered aircraft in October 1907, he flew his Santos-Dumont 14-*bis* standing up. Grafted into the middle of the odd machine's fuselage was what looked like a wicker balloon basket. It was obvious Santos-Dumont had come to heavier-than-air flight from balloons and dirigibles.

Common sense generally prevailed, however, and most first-generation aviators, successful or not, rightly assumed a person should sit down to fly. But even in the seminal year 1909, when successful flying really took off in the wake of Wilbur's demonstrations at Le Mans the previous year, this was where agreement ended.

For example, Louis Blériot sat in his Blériot XI like an automobile driver, whereas Hubert Latham sat atop his Antoinette as if crewing a rowboat. And when Santos-Dumont finally came up with a working airplane, he huddled beneath the wing and engine of his Demoiselle. So low was he seated that his gloved hands served as wheel brakes.

Similar confusion attended the flight controls. The Wrights at different times used the motion of pulling back or pushing forward on a lever to lower the nose of their gliders (the latter, being intuitive, won out). As late as 1908, the Wright Model A—their first airplane produced in number—had two different control systems, one reflecting Orville's preference for rudder actuation and the other Wilbur's.

Some early airplanes sported control wheels while others had sticks. The Antoinette had fore-and-aft control wheels mounted perpendicular to the direction of flight at the sides of its cockpit. A pilot would control the ailerons by rolling the left-hand wheel forward or backward and the elevator by doing the same with the wheel at right. As for the rudder, a pivoting foot bar controlled that.

The Blériot XI had a wheel on top of a control stick, but this wheel didn't turn. Instead it simply served as a round grip. The elevator was actuated by moving this wheeled stick forward or aft, while wing warping was controlled by side-to-side movements. The stick could be moved to any quadrant in a circle for combined inputs.

This excellent idea was actually developed independently by Robert Esnault-Pelterie and Louis Blériot. It is called the *cloche* (French for "bell") control system. The name refers to the shape of the bottom end of the control stick below the universal joint allowing it to pivot. This bottom end splayed like a bell where the airplane's elevator and aileron or wing-warping control cables attached to it.

As for control systems using wheels, the Deperdussin company

Hubert Latham directed the Antoinette in flight using fore-and-aft control wheels at each side of the cockpit.

came up with the winner. Drawing on maritime practice, Deperdussin pilots simply turned a control wheel in the direction they wished to bank. To this intuitive lateral control, Deperdussin added fore-and-aft wheel movement providing equally intuitive pitch control via the elevator.

The Deperdussin, Antoinette, Blériot XI, Farman, and almost all other early airplanes used a pivoting foot bar to control the yaw axis. The "rudder bar" idea caught on quickly because it was so easy to wire up and intuitive to use. To yaw the airplane's nose, one had only to push with the foot on the side to which you wished the nose to go. The rudder bar also facilitated smooth turns, which require coordinated application of yaw and roll control.

After World War I, rudder bars gradually gave way to today's rudder pedals. Although they function identically, pedals have the advantages of greater comfort during maneuvers, since one's feet are less likely to slip. Rudder pedals also accommodate wheel brakes, actuated by depressing the tips of the pedals with one's toes.

Interestingly, the Wrights gave up foot controls after their 1901 Glider. In their 1905 and 1908 Flyers, all three axes were controlled independently by the hands, which also managed engine throttling. Glenn Curtiss likewise went his own way. The *June Bug* and Model

D Pushers had a wheel on a movable control column. As with most airplanes, pulling the wheel toward you brought the nose up, while pushing it away put the airplane into a dive. However, turning the wheel laterally worked the rudder to yaw the airplane's nose left or right. Roll was controlled by a shoulder-operated yoke that worked the ailerons. To bank, one simply tilted one's body to the desired side.

All this was potentially very confusing. It lasted until World War I, which drove consensus because large numbers of pilots needed to be trained. Once sent off to the front, moreover, they had to be able to fly whatever airplanes they found there and to transition to newer equipment as it arrived.

Cockpit instruments likewise evolved haphazardly. These fall into two basic classes: flight instruments and engine instruments. The former came first and had the humblest of births at the hands of the Wrights at Huffman Prairie. As aviation historian and Wright biographer Tom Crouch recounts:

> The brothers had trouble orienting themselves in turns, frequently miscalculating, banking too steeply, or allowing the nose to rise so high that the aircraft stalled. The answer was a long string tied to the crossbar of the elevator. When the craft was flying straight and level, the string blew directly back toward the pilot. When banking, or flying with the nose up or down, the position of the string enabled him to gauge the attitude of his machine.[1]

The first real instrument was probably the compass. Once pilots began flying any distance, they needed to know where they were headed. Otherwise, fliers before World War I flew only by external cues and the feel of the airplane. That changed in the war, of course, because military pilots needed to know how high they were, how fast they were flying, how much fuel they had left, and how healthy their engines were.

Although World War I stuffed more instruments in the cockpit, it did not drive consensus. Instead, it was pretty much up to each designer to place these instruments at his whim. A generation later, World War II finally standardized the layout of cockpit instruments.

In between these wars, a quiet revolution occurred in aviation: human beings taught themselves to fly at night and in bad weather. This singular achievement took place near New York City even as the stock market crashed on Wall Street in 1929, triggering the Great Depression. More specifically, it happened on the Hempstead Plains, a great treeless expanse of Long Island that was one of the few natural grassland prairies on the North American continent's eastern seaboard. Often called the U.S. cradle of aviation, these plains were home to many of the nation's most active early flying fields.

Weary of upstate New York's heavy snowfalls, Glenn Curtiss relocated his airplane factory to the south end of the state, settling on the Hempstead Plains in July 1909. After testing his *Golden Flyer*, Curtiss shipped it to Reims, France, and flew it to victory that August to claim the Gordon Bennett Cup. In turn, New York hosted many of the same aviators the next year in a major air meet at Belmont Park, the horse racetrack at the western edge of the Plains.

During World War I, the United States trained thousands of Army aviators at two military airfields on the Plains. One was Roosevelt Field. Bordering Curtiss Field immediately to the east, it was named in honor of Lt. Quentin Roosevelt, the former president's youngest son, who died valiantly in aerial combat over France in July 1918.

The other military installation lay to the south. This was Mitchel Field, named for John Purroy Mitchel, once New York City's youngest mayor, who had died in a flight training accident in Louisiana in 1918. Charged with New York's aerial defense, Mitchel Field in the late 1920s was one of the U.S. Army Air Corps' premier installations.

Crickets chirped in the surrounding night. Inside a dark hangar, feet scuffed concrete. Shadows swung as a handheld lantern, the building's sole source of light, was handed up to a man crouched on the lower wing of a silver-and-yellow biplane. He lowered the lantern onto the seat of the rear cockpit and closed an improvised canvas hood over it, plunging the Mitchel Field hangar into darkness.

Jimmy Doolittle and his Full Flight Laboratory team stood in a circle around this airplane, a Navy Consolidated NY-2 selected for its stability. They scrutinized the trainer from every angle. Wherever they saw light escaping through chinks in its fabric, they pointed it out to a mechanic, who hurried forward with a can of aircraft dope and opaque fabric patches. With a few swipes of his dope brush, the gleam vanished.

Within minutes, all traces of the lantern's glow had been eliminated. Doolittle, a thirty-two-year-old Army first lieutenant, was satisfied. When he flew by reference to instruments alone, no light would enter from outside to provide clues as to his airplane's orientation.

In the latter 1920s, the world knew how to build rugged, capable airplanes, but bad weather often kept these fine machines on the ground. Worse still, it often caught them aloft and claimed lives when pilots were denied a view of the ground.

Fog was the killer most feared, but it was just one of many. Depending on where in the world people flew, they might also contend with thunderstorms, hurricanes, tropical squalls, monsoons, blizzards, typhoons, and sandstorms. Even clear skies and calm winds were dangerous for airplanes overtaken by falling darkness on a moonless night.

Airmail pilots in the 1920s knew these dangers all too well. Flying day or night in open-cockpit machines with unreliable engines, they took in stride whatever nature dealt them. Those who survived did so by knowing that, contrary to the prevailing belief, there was no such thing as seat-of-the-pants flying.

Mail pilots occasionally found themselves trapped above a solid undercast. When that happened, their orders were to grab the mailbags and bail out. Lindbergh did that twice because he knew descending through clouds was a sucker's bet. Even if the clouds stopped short of the ground, there was no guarantee you could keep your airplane upright long enough to break safely out in the clear.

Physiological tests confirmed what these aviators knew from bitter experience: human beings, when deprived of visual cues, cannot maintain their spatial orientation. To prove it, researchers strapped blindfolded volunteers into gimbaled chairs and set them spinning and tilting head over heels at the same time. Within seconds, the volunteers were hopelessly wrong about what was happening to them in terms of motions, accelerations, and orientation.

The culprit is the human vestibular system, which provides our sense of balance. Easily confused when motion sets fluids swirling in our inner ears, and absent visual cues to provide a fixed frame of reference, it is a highly persuasive liar.

Harry Guggenheim was born to great wealth in 1890. The son of New York–based mining magnate Daniel Guggenheim and Florence Guggenheim, Harry attended Yale University in Connecticut, worked briefly in the family businesses in Mexico and elsewhere, and completed his education at Cambridge, England.

This multimillionaire accustomed to privilege was in his late twenties as World War I raged in Europe. Realizing his country would inevitably become involved and determined to be of help, he purchased a Curtiss flying boat and took flying lessons. When the United States entered the war in April 1917, he joined the U.S. Navy and served with distinction as a naval aviator in France, England, and Italy.

Like his parents, young Harry possessed a strong sense of civic responsibility. He was devoted to aviation, a fascination shared by his father. Consequently, when the elder Guggenheims formed a philanthropic foundation in 1924, part of its endowment was earmarked to benefit the emerging science of flight.

Perceiving a critical lack of U.S. expertise, the Guggenheim Foundation in 1925 established an aeronautical engineering program at New York University (NYU). That was just the beginning; the following June the family chartered the subordinate Daniel Guggenheim Fund for the Promotion of Aeronautics, whose eminent board of trustees worked closely with interested U.S. government agencies. Harry served as president of the Guggenheim Fund, which in four short years of existence (1926–30) revitalized aviation in the United States.

The U.S. aviation scene had been in the doldrums since World War I, with America lagging ever farther behind Europe. Under Harry's gifted leadership, the Guggenheim Fund worked to reverse this trend through infusions of capital focused on several vital fronts.

One was education. Based on the successful NYU effort, the Guggenheim Fund also endowed aviation programs at Stanford, Harvard, the University of Washington, Northwest, MIT, the University of Michigan, and several other schools. One of these new programs—the California Institute of Technology's Guggenheim Aeronautical Laboratory (GALCIT)—scored a coup by luring Hungarian aerodynamicist Dr. Theodore von Kármán, a brilliant former student of Ludwig Prandtl, away from Germany to serve as its director.

The Guggenheim Fund also fostered aviation research and development, sponsored safety-plane competitions, pioneered improved aviation weather forecasting, and in general helped U.S. commercial aviation make the transition from carrying the nation's airmail to transporting passengers.

In the midst of all this activity, a twenty-five-year-old airmail pilot named Lindbergh arrived at Curtiss Field with the intention of flying nonstop to Paris. It was the spring of 1927 and he was the last of many to gather at the Hempstead Plains to try for the Orteig Prize, a $25,000 purse posted by wealthy hotel owner Raymond Orteig for the first airplane to fly nonstop between New York and Paris.

Curtiss Field and neighboring Roosevelt Field, whose longer runway Lindbergh would use, were not far from Harry Guggenheim's palatial estate on Long Island's North Shore. Shortly before the young aviator took off on May 20, Guggenheim stopped by to wish him well. "When you get back to the United States," he said gamely, "come up to the Fund and see me."[2]

Nevertheless, he privately felt that the slender youth a dozen years his junior stood little chance. How could even the ablest flier hope to conquer that vast ocean in a frail single-engine monoplane? As a frequent ocean voyager and former Great War flying-boat pilot, Guggenheim knew all too well the fearsome power of the North Atlantic's waves and weather.

"This fellow will never make it," he told himself, fearing the worst. "He's doomed."[3]

But Lindbergh *did* make it, and the world went crazy. Overnight, it seemed, Americans were passionate about flight, which claimed center stage in the cultural mainstream. Aviators became the nation's heroes and flight its future. Reflecting this euphoria, Wall Street opened the floodgates to indiscriminate investment in companies building or operating airplanes.

Lindbergh did look Harry Guggenheim up on his return, and the two became fast friends. The philanthropist generously opened his mansion to the youth, granting him solitude from his sudden celebrity, and sent him to his tailor for a tuxedo, tails, and everything else that Lindbergh's sudden rise to prominence demanded.

Charles Lindbergh's flight evoked tantalizing visions of commercial air travel. In one day, the possibility of intercontinental air travel went from wildly fanciful to probable. Crowning the Roaring Twenties—a frenetic era of jazz, flappers, Art Deco design, social upheaval, and burgeoning industry—this newfound human mobility was nothing short of intoxicating.

All the technological pieces were now in place for this dream to come true. All, that is, but one. The inability of airplanes to operate in darkness or bad weather meant that air travelers were held hostage by the elements, their flights frequently delayed or canceled. So long as this limitation existed, regularly scheduled air services would be a chimera. The military too chafed under the constraints of bad weather, which prevented its ablest machines and most capable pilots from completing assigned missions.

Harry Guggenheim knew all about it. When he began flight training early in 1917, his instructor pointed to his Curtiss flying boat's sparse panel. "See those instruments?" the man said. "Pay no attention to them. In the first place, they are not accurate and I want you to get the feel of the ship regardless of instruments."[4]

Back then most aviators felt they could trust their instincts. Guggenheim was doing just that when he emerged from broken overcast above Long Island Sound not long afterward. He wrote later at being "amazed on coming out of the cloud to see a ferryboat below apparently tipped at an angle of about forty-five degrees and rapidly turning around in short circles."[5]

In fact, it was he who was tilted and circling despite an utter conviction of continued straight-and-level flight during his foray into the clouds. That sobering memory drove him now to address the problem of blind flying. At that time, many countries were working on it, but those efforts were getting nowhere. To succeed, the Guggenheim Fund for the Promotion of Aeronautics needed to assemble the best possible team. But who should lead it?

Guggenheim put this question to Captain Emory Land, a naval officer serving as his second in command on the fund's board. Risking the ire of his own military service, Land enthusiastically recommended Army first lieutenant Jimmy Doolittle.

Doolittle had won the 1925 Schneider Trophy Race—an international competition for float planes—ahead of two U.S. Navy fliers, making better water landings to boot. More to the point, this Army flier had earned a master's degree and a doctorate in aeronautical science in 1924 and 1925, respectively, from MIT. No one else combined his skills in the air with such solid academic credentials. If anyone could solve blind flight's baffling challenges and banish its demons, it was Doolittle.

Harry Guggenheim agreed. Setting the wheels in motion in 1928, he collaborated with all interested parties to establish the Full Flight Laboratory at Mitchel Field. At his request, the Air Corps detached Doolittle to the Guggenheim Fund to serve as the laboratory's director. In addition to an expert staff and equipment, the fund had two Navy biplanes. One was a fast Vought O2U-1 Corsair for Doolittle's use in coordinating with suppliers. The other was a Consolidated NY-2 Husky, a stable training plane that took off and landed slowly. Equipped with heavy-duty wheels instead of the usual floats, it was an ideal platform for prototyping a new way to fly.

Born in Alameda, California, on December 14, 1896, James Harold Doolittle spent his formative years in Nome, the raw Alaskan town to which his father had drifted following the Klondike Gold Rush's northward lure. Nome's hardy, rough-and-tumble lifestyle instilled in the boy a strong sense of fair play and self-reliance. Frontier life also taught him how to use his fists.

Doolittle's mother brought him back to California, settling in Los Angeles. There he attended the first U.S. air meet, which was held at

Dominguez Field at the start of 1910. Later Doolittle built an unsuccessful glider and was halfway through constructing a Santos-Dumont Demoiselle monoplane in 1912 when a windstorm put an end to his early aerial aspirations.

Jimmy Doolittle attended high school in Los Angeles, where one of his classmates was legendary film director Frank Capra. Short and compact, Doolittle excelled at boxing and tumbling. Marrying his high school sweetheart, he studied mining engineering in college until World War I interrupted his studies.

Doolittle volunteered for the Army Air Service. From the outset of training, he proved such a good pilot that, to his frustration, he was kept stateside to train others rather than going overseas into combat. In 1924, he flew across the United States in less than twenty-four hours. In 1925, he won the Schneider Trophy Race. Two years later he performed what is believed to be aviation's first outside loop.

Archie League, history's first air traffic controller, evaluates a searchlight and experimental radio equipment in the battle against fog.

At Mitchel Field, Jimmy Doolittle began by defining the missing instrument capabilities that would allow "safe and reliable flights despite weather conditions," as he put it.[6] At his disposal during all this brainstorming were experts from the Radio Frequency Laboratory and Bell Labs, MIT, the U.S. Department of Commerce's Aeronautics Branch and Bureau of Standards, and various private companies in the New York area.

Chief among these was the Sperry Gyroscope Company of Brooklyn, New York. Doolittle met with Elmer Sperry Sr., a noted U.S. inventor in his late sixties, to discuss a new flight instrument he wanted developed. As sketched out by Doolittle, this instrument's round face had a peripheral ring marked like a compass rose surrounding a central depiction of the earth's curving horizon. The ring would rotate to display the airplane's heading while the central horizon would indicate its attitude in relation to the ground.

Sperry's earlier inventions ranged from electric streetcars to arc lamps, but he specialized in gyroscope applications. What really put his company on the map was his gyroscopic compass for guiding ships, whose pitching motions and steel hulls made navigating by a conventional magnetic compass problematic.

Sperry assured Doolittle his company could develop the requested instrument, but he proposed doing it as two separate instruments for ease of manufacture. One would be the *gyroscopic compass* (later simply called *gyro compass*) and the other the *artificial horizon* (today *attitude indicator*).

The gyro compass was needed because a floating magnetic compass, while fine for straight-and-level flight, oscillates and precesses badly when the airplane turns, climbs, or descends. It is also affected by changes in acceleration. The result is constant bobbing and spinning that renders a floating compass virtually useless during the maneuvering that would be required for instrument approaches and landings.

Equally critical to Doolittle's mission was the artificial horizon, an instrument that shows an airplane's orientation relative to the earth's surface. Even if clouds or darkness hid the ground, it would tell the pilot whether the airplane's nose was pointing above the horizon, in line with it, or below. It would also show whether the wings were level or tilting to one side or another.

The senior Sperry assigned his son, Elmer junior, to the project in a wholehearted commitment that saw the latter actually join Doolittle's team at Mitchel Field. Father and son had a special reason for supporting this effort so fully. Another of Elmer's sons—dashing Lawrence Sperry, a pioneer flier and prolific inventor who built the world's first aviation autopilot in 1914—had been working along similar lines when he disappeared over the English Channel on December 23, 1923.

His Sperry Messenger, a small biplane of Lawrence's own design, was found soon after floating in the water, its cockpit empty. Closure finally came for the grief-stricken family on January 11 when his body too was recovered.

Lawrence Sperry had taken off in fog.

During World War I, Allied aviators were issued wristwatches. A valuable piece of equipment, the wristwatch allowed time checks even when one's hands were busy on the controls. Previously regarded as inferior to the pocket watch and an item generally of women's apparel, the strap-on timepiece's popularity with returning veterans brought it into mainstream use by all.

In the latter 1920s, a shy young German American living in New York City went looking for local wristwatch makers whose skills could help him build a new aircraft instrument he had conceived. It was an altimeter at least ten times as accurate as the World War I–era altimeters then in use.

Born in Freudenstadt in 1900, Paul Kollsman trained as a civil engineer in Stuttgart and Munich before immigrating to the United States. Arriving in New York at age twenty-three, he worked as a truck driver's assistant before finding a better job building aviation instruments for the Pioneer Instrument Company of Brooklyn.

Excited by his idea of a truly accurate altimeter, Kollsman left Pioneer in 1928 to start his own company. It began with an initial capitalization of $500 in the garage behind his house in Brooklyn. He then made the rounds of New York City–based watchmakers, for they alone possessed the manufacturing precision and repeatability required to make his design's miniaturized gears and other precision components.

One finds everything in New York. Kollsman located Swiss craftsmen who could build and repair the finest wristwatches. An agreement was concluded to apply the skills of one area of human endeavor to benefit another. Parts in hand, Kollsman carefully built working prototypes of the *sensitive altimeter*, as he called it.

Kollsman was ahead of his time. There was no market for his device because aviation was then doing fine with World War I–style altimeters. Although they were inaccurate and indicated differently from one day to the next, those existing devices were fine for telling pilots roughly how high they were in the early days before instrument flying existed.

World War I altimeters had large-diameter faces and a somewhat nautical look. Made of brass or steel with heavy mounting flanges and thick glass faces, they displayed the airplane's range of operating altitudes in 1,000-foot increments. A single needle set against this closely spaced scale showed the airplane's approximate height.

The reason that World War I altimeters were so capricious is of course that ambient air pressure changes according to the weather. Balmy high-pressure days made them read too low, whereas stormy conditions gave false indications on the high side.

Kollsman's sensitive altimeter differed greatly. For starters, it was designed to be accurate within 20 feet or less of actual altitude. Smaller in diameter, this instrument had two indicator needles set against a clockwise scale that read from 0 to 9. Each increment denoted 100 feet, not 1,000 feet. As the airplane climbed, this altimeter's big needle swept clockwise around and around these increments. Each time it passed 0 again, the smaller needle advanced to indicate an additional 1,000 feet of altitude.

The Kollsman altimeter thus resembled a clock numbered to 10, which made it easy and intuitive to read. If the big hand was on the

7 and the little hand on the 2, the airplane was flying at 2,700 feet. A third indicator needle could be added to display altitudes above 10,000 feet.

Unlike a barometer on a wall, an aviation altimeter should indicate identically regardless of the day's weather conditions. Consequently, Kollsman made his altimeter adjustable to compensate for ambient air pressure changes. Using a setting knob, pilots could simply dial the current barometric pressure into an inset window to ensure that the device read accurately. It also worked the other way around; setting the altimeter to show the airfield's known elevation while on the ground revealed the current barometric pressure in the setting window.

On cross-country flights, pilots of the future would receive by radio the destination airport's current barometric setting and enter it in their altimeters before landing to ensure accurate readings. Of course, this capability was still many years off. While some airlines and military air services had voice communications technology, its use remained experimental at the end of the 1920s. Years would pass before it became broadly available as a vital adjunct to routine instrument flying, which likewise did not yet exist.

Kollsman was thus a visionary. Had he been forced to wait until the world caught up with his ideas, the company he founded would have gone out of business. Fortunately for him, word of his invention reached the ears of a scientific team at Mitchel Field, less than 15 miles (25 kilometers) away.

After meeting Kollsman and reviewing his invention, Jimmy Doolittle took the young German American aloft several times. The young man sat in the open front cockpit ahead of Doolittle, his sensitive altimeter mounted on a board on his lap. The device did all that he claimed, demonstrating uncanny accuracy landing after landing.

Existing altimeters were no good for blind landings, but this one was. Fortuitously, the Full Flight Laboratory team had stumbled across it all but gift-wrapped, an exquisite piece of technology cobbled together by a serious youth still in his twenties.

Doolittle now had cockpit displays allowing flight control more precise than ever before possible without external visual reference cues. In addition to Kollsman's altimeter and Sperry's two new devices, the airplane had an existing gyro instrument called a turn and bank indicator as well as airspeed and rate-of-climb indicators.

These flight instruments were half the blind-flying equation. Operating under the hood, Doolittle would also need to know where he was in the sky, and that required *radio-navigational aids*. To meet this challenge, Doolittle flew the NY-2 to Boonton, New Jersey, where technicians at the Radio Frequency Laboratory installed two additional collaboratively defined and developed electronic devices in the trainer's ever more crowded rear cockpit.

The first was a homing indicator that would let Doolittle follow an electronic beam projected along the runway's extended centerline (today called a *localizer*). The primitive device in the NY-2 had two vibrating reeds at each side of its display window. If one reed vibrated faster than the other, the airplane was off to that side; when both reeds vibrated at the same rate, it was in the middle, aligned with the runway.

The second electronic aid was a fan beacon with a vibrating-reed display. When Doolittle passed over this electronic marker, the vibrating would stop, letting him know to throttle back for a final descent at a constant shallow rate to the runway. Landing this way meant flying the ship literally into the ground, but the NY-2's beefed-up landing gear was built to take it.

There was no third beam to tell Doolittle whether he was on the proper glide slope. That would come later.

Jimmy Doolittle, arguably the greatest pilot of all.

Laboratory business took Jimmy Doolittle to Buffalo one late winter day. Returning at night in the Vought O2U-1 on March 15, 1929, he headed south into deteriorating weather. To avoid unseen mountains, he planned to fly via Rochester, Syracuse, Utica, Schenectady, Albany, and down the Hudson River Valley.

On the west side of the Hudson, Doolittle slowed up to pace the lit windows of a southbound train. It entered a ravine, forcing him to turn away. He considered landing at West Point's parade ground, but conditions were just good enough to press southward. The lights of New York City welcomed him, but turning east around Battery Park toward Mitchel Field, he found the East River and all points south firmly socked in.

Governor's Island was shrouded in fog, and so was the Yonkers Golf Course back up the Hudson. Doolittle started to land at Battery Park only to have a running figure wave him off. He couldn't land there for fear of hurting the man.

The George Washington Bridge, built between 1927 and 1931, was half completed at that time. Doolittle flew repeatedly past its great vertical pilings not knowing they were there. By now it looked to him as if he would have to ditch in the river. He undid his parachute harness to be ready to swim for it, but on closer inspection the water looked decidedly uninviting.

He climbed again and headed westward through thickening fog, hoping to land at Newark Airport, but it too was socked in. With the gas gauge hovering on empty, he climbed to a thousand feet and broke out into a still, starry night. The only thing to do was to fly westward to a less populated area and hit the silk.

About this time he realized his parachute harness was undone. As he refastened it, a rotating beacon shining through a hole in the mists below caught his eye. It might denote an airport, and the dark area next to it might be the flying field. Turning on his landing light, he dove through the hole to investigate.

A treetop tore through the left wing. The Corsair still flew but was just about out of gas. Picking his best option in the scything landing light, Doolittle set down, purposely wrapping his left wings around a tree to take the impact.

The broken airplane's fuselage came to rest on its side, and Doolittle emerged without a scratch. Fog had missed claiming another aviator.

On September 24, the Full Flight Laboratory team awoke to dense fog. It draped the field like a blanket, reducing the ceiling and visibility to zero. For Doolittle, it was the answer to a prayer.

"I decided to make a real fog flight," he later recounted. "The NY-2 was pushed out of the hangar and warmed up. The ground radios were manned and the radio beacons turned on. I taxied out to the middle of the field and took off. Coming through the fog at about 500 feet and making a wide swing, I came around into landing position. By the time I landed 10 minutes after takeoff, the fog had started to lift."[7]

History's impromptu first blind flight was logged with no witnesses beyond the immediate team. Already summoned, Harry Guggenheim and other observers arrived shortly thereafter to watch an official re-

Doolittle performed the world's first blind flight on September 24, 1929, using instruments and procedures he had helped devise.

staging of the event. Because the fog had now cleared and the airfield was again open, Guggenheim ordered Ben Kelsey—a young Army second lieutenant assigned as the project's second pilot—to ride along in the front cockpit. Kelsey would take over only if another airplane showed up, spoiling the attempt; otherwise he would hold his hands high in the air to show Doolittle was at the controls.

The hood was again closed over Jimmy Doolittle. He taxied out and turned onto the beam-defined runway. Gunning the engine, he climbed straight ahead to a thousand feet, performed a standard-rate turn with the aid of a stopwatch, flew a timed downwind leg, and turned back toward the runway.

Descending to 200 feet, Doolittle leveled out and held that altitude until his instruments showed he had passed over the fan marker. At that beacon signal, he brought the throttle back to a mark inscribed on the throttle quadrant. Long practice working out the procedures

Flying under a hood, Doolittle took off, flew a rectangular pattern, and landed again without ever once seeing the ground.

had told him that this combination of attitude and engine power setting yielded a gradual 200-foot-per-minute descent to the runway.

In contrast to his earlier approach, this one ended with a bounce—Doolittle called the landing "sloppy"—but they were down. It was a momentous occasion. The integrated technical capabilities demonstrated by Doolittle's team would forever change aviation.

"Fog Peril Overcome," trumpeted the front page of the *New York Times* the following morning. "Man's greatest enemy in the air, fog, was conquered yesterday at Mitchel Field when Lieutenant James H. Doolittle took off, flew over a fifteen-mile course, and landed again without seeing the ground or any part of his plane but the illuminated instrument board." Rightly heralding the dawn of a new era in flight, the *Times* observed that "aviation had perhaps taken its greatest single step in safety."[8]

Doolittle went on to set more records and win the Thompson and Bendix trophy races in the early 1930s, making him the only pilot ever to win all three of the major U.S. air races between the wars. After leading the daring Doolittle Raid four months after Pearl Harbor, for which he received the Medal of Honor, he held key commands and finished the war a lieutenant general. Granted a fourth star in retirement, Doolittle died in 1993 at age ninety-six.

Perhaps the greatest pilot who ever lived, Doolittle always considered leading the Guggenheim Fund's Full Flight Laboratory to be his most significant accomplishment.

When the ten-passenger Boeing 247 airliner entered service in 1933, military pilots looked on enviously. Whereas they flew slow, drafty open-cockpit biplanes, United Air Lines pilots—the only ones to get their hands on the sleek Boeing in its first year of commercial service—flew at 185 mph (300 km/h) in the heated comfort of a fully enclosed cockpit.

Unlike military airplanes of the day, the Boeing 247's instrument panel boasted the gyro instruments and radio gear that let airline flights operate in bad weather. Using the nation's newly developed four-course radio range system and instrument flight procedures, Model 247 flight crews following primitive electronic guidance sig-

The center of the Boeing 247 instrument panel had three gyro instruments for blind flying: a directional gyro (top left), artificial horizon (top right), and turn-and-bank indicator (middle).

nals in their headphones navigated airways and performed instrument approaches to properly equipped airports.

It was all highly imprecise and subject to interference. Even so, it was nothing short of transformative—aviation's equivalent of having ships that could go anywhere versus ones that must hug the coastline.

Boeing was so proud of the Model 247's instrument panel that it had a female employee model it on her lap. Featured at top center were a gyro compass and artificial horizon developed under the auspices of the Guggenheim Fund's Full Flight Laboratory at Mitchel Field. The instrument panels of the next three Boeing airliners would also mark aviation milestones.

In 1940, the pressurized Boeing 307 Stratoliner entered international service with Pan Am and domestic U.S. service with TWA. The Stratoliner had so many switches, gauges, and controls competing for space on its instrument panel that Boeing engineers relocated a lot of them to a separate panel on the right-side wall behind the copilot.

The Stratoliner thus became the first production airliner designed

The copilot and flight engineer confer aboard a Boeing 307 Stratoliner in 1940.

with a third flight crew position, for the *flight engineer.* Dornier in Germany actually hit upon this idea first, but the company's unsuccessful Do X flying boat of 1929—which needed a flight engineer to manage its twelve engines—did not achieve production.

The flight engineer's job on the Stratoliner was to manage the airplane's engines, fuel, pressurization, cabin temperature, electrical, and other systems to free the pilot and copilot to concentrate on the flying. Large commercial airliners would have three-person flight crews for decades to come.

This is not to say that airliners never had three or more people in the cockpit before. Before World War II, many long-range transports carried dedicated navigators and radio operators. However, those peripheral cockpit positions were not central to the basic operation of the airplane or its systems.

The Boeing 314 Clipper, last and largest of Pan American Airways' ocean-spanning flying boats of the 1930s, had perhaps the biggest cockpit of any airplane. Behind the pilot and copilot was an area grand enough to be the bridge of a ship. Arrayed around this open space were stations for a radio operator, a navigator, and a flight engineer. Although the idea arose with the Model 307 Stratoliner, the Model 314 Clipper actually entered service first.

Boeing 314 Clipper crews had ample space to stand up, walk around, study large ocean navigation charts spread across a plotting table, shoot the stars or sun with a celestial navigation octant, and so on. At night, blackout curtains cordoned off the pilots from the rest of the crew area so that the aft cockpit's lights would not compromise their vision.

As if that weren't enough, a door at the rear led to sleeping quarters for a relief crew because flights could last upward of twenty hours. All of this existed above the Clipper's passenger areas, an idea Boeing later revisited in the design of its 747 jetliner.

World War II profoundly changed the aviation scene. Wartime funding and urgency again accelerated the development of flight technologies, creating larger and more capable transport planes.

After the war, Boeing brought out the Model 377 Stratocruiser. Blunt and fast, this barrel-chested airliner of the late 1940s embodied technology developed for the Boeing B-29 Superfortress bomber, the most sophisticated airplane of World War II. The Boeing Stratocruiser would be the largest, heaviest, and most powerful piston-engine airliner of all time, and it too featured a noteworthy cockpit.

When the Stratocruiser came into being, virtually everyone who crewed it had prior experience on the ubiquitous but far smaller Douglas DC-3 airliner or its military cousins. In wet weather, the pilots of those famous Douglas transports often donned rain gear because the cockpit windows tended to leak. If those fliers thought that stepping up to the pressurized and presumably leakproof Stratocruiser would keep them dry, they were sadly mistaken.

In high-altitude cruise, moisture in the cabin air gradually condensed against the inside of the airplane's cold metal skin. As its nose pitched down on descent, this moisture slid forward to rain down on both pilots. Equally miserable was time spent on the ground at hot airports because the Stratocruiser's heavily glazed cockpit mercilessly exposed its crew to the sun. It was ironic that an airplane so comfortable for the passengers could so torture its crew.

But where the Boeing 377 cockpit truly stands out is in its sheer complexity. Hands down, it placed heavier demands on flight crews than any other airliner in history. Fortunately, air transport pilots

Introduced at the end of the 1940s, the Boeing 377 Stratocruiser was the largest and most powerful of the great piston airliners that flew after World War II.

would never again have to contend with so many gauges, levers, switches, dials, procedures, and checklists because technology was about to give this "aerial office" a makeover.

This relief came as the by-product of the commercial jet age. It would be the first of two major leaps toward greater flight deck simplicity.

Germany and Great Britain both fielded operational military jet aircraft during World War II. After the war, Britain was the first to apply turbine propulsion in the commercial arena with its pioneering but flawed de Havilland Comet, a thirty-six-passenger airliner that entered service in May 1952. More successful were the Tupolev Tu 104, Boeing 707, Sud Aviation Caravelle, and Douglas DC-8 jetliners that arrived by decade's end. Also, starting in the 1950s, turboprops—turbine engines driving propellers—entered service on shorter routes.

During the next decade, Britain introduced the Trident, VC-10, and BAC One-Eleven jet transports while the United States gave airlines the 727, DC-9, and 737. By the end of the 1960s, routes short and long were being flown by a variety of jets.

Stratocruiser crews contended with the most complex cockpit of any airliner.

Jets were easier to fly from the flight crew's perspective because eliminating piston engines and propellers also dispensed with a lot of complexity. For example, to change engine power settings on a DC-7, Douglas' last propeller airliner, the flight crew had managed four controls per engine: throttle, mixture, propeller rpm, and boost (supercharging). In contrast, the DC-8 and other jetliners have just one thrust lever per engine.

This was a step in the right direction, but crews still contended with the need to scan and interpret a bewildering number of instruments. They were being overwhelmed by the sheer amount of information presented by the round-faced electromechanical instruments then in use. Addressing this challenge, NASA led research aimed at processing this information and presenting it to flight crews on video screens. This industry-government collaboration defined effective graphic interfaces that organized the information and prioritized what the crew needed to know according to the immediate situation, with additional information readily available as desired.

The result was the modern *glass cockpit*, so called because it employs *electronic flight instrument system* (EFIS) screens to provide the pilot and copilot with the primary flight and navigation information formerly supplied by a variety of individual instruments. Yet more screens in the center of the instrument panel display engine information and perform crew-alerting functions.

EFIS debuted in airline service with the Boeing 767 in 1982 and Airbus A310 the following year. Universally adopted and often retrofitted to older jets, this display technology in conjunction with sophisticated flight management computers has revolutionized airline operations. As a result, modern flight decks have far fewer instruments and controls than earlier-generation flight decks.

Starting in the early 1980s, EFIS and flight-management computers so effectively reduced flight crew workloads that even the biggest jets no longer needed a flight engineer. Today glass cockpits are ubiquitous. Even the Airbus A380—the world's largest commercial airliner—has a two-person flight deck, although its spacious cockpit also offers three jump seats.

Improvements continue apace as technology advances. One exam-

Modern jetliners like the Airbus A380 rely on computerized "glass cockpits," so named for their instrument display screens.

ple is the adoption of flat-panel liquid-crystal display (LCD) screens that are larger, easier to read, take up less room, use less power, generate less heat, last longer, and are more reliable than the cathode ray tube (CRT) displays of first-generation EFIS. Whereas CRTs—which are like the picture tubes of old-fashioned television sets—wash out when direct sunlight splashes on them, LCD screens remain easy to read.

Farther down the line, these advanced screens may in turn give way to organic light-emitting diode (OLED) screens. Now being pursued by the consumer electronics industry, OLED displays promise to be still lighter in weight, sharper, and more colorful, and to use much less power. They will also be thinner—perhaps as little as a few millimeters—because they do not require internal backlighting, as do LCDs.

Ongoing human-factors research continues to refine the modern flight deck. Specially trained scientist-engineers focus this fascinating field's expertise on how pilots interact with the airplane and one another, as well as how the airplane communicates with its human pilots. The result is ongoing improvements in the human-machine interface and crew-coordination procedures.

Human-factors engineering also contributes to a disciplined, highly beneficial methodology known as *crew resource management* that continues to make aviation safer. One example of CRM at work is the sterile cockpit rule, a regulation implemented in the early 1980s that prohibits extraneous conversation when the airplane is below 10,000 feet. Flight crews can chit-chat during high-altitude cruise, but not during the more demanding phases of flight.

Some of the technologies found in many modern flight decks were first developed for military use. In 1988, for example, Airbus audaciously introduced fly-by-wire to commercial aviation with its single-aisle A320 jetliner. Developed for military fighter jets, fly-by-wire flight control systems eliminate the physical connection between the flight crew's controls and the control surfaces they actuate.

Fly-by-wire saves weight because the cables and pushrods that once physically deflected the airplane's control surfaces have been replaced by electronic signals controlling actuators at those surfaces. Depending on the programming of the airplane's flight-control computers, fly-by-wire can also significantly alter or improve an airplane's flight characteristics.

What would Hubert Latham—seated atop his Antoinette like a fisherman in a rowboat—have made of the modern flight deck? One can only imagine.

10 AERO PROPULSION

PROMETHEUS IS PUSHING

Air surrounds me as water surrounds the submarine boat, and in it my propellers act like the screws of a steamer.

—JULES VERNE, *ROBUR LE CONQUÉRANT*, 1886

Two centuries ago, George Cayley predicted human beings would fly once a first mover became available that generated "more power in a given time, in proportion to its weight, than the animal system of muscles."[1] By "first mover," he of course meant an engine.

Steam power became available early in the nineteenth century. Although many flight experimenters looked to it hopefully, the low power-to-weight ratio of steam engines rendered them generally unsuitable to aviation use. Instead, it would be the internal-combustion engine powered by gasoline that would see Cayley's prediction realized.

Belgian-born French inventor Jean-Joseph-Étienne Lenoir created the world's first internal-combustion engine in 1859. Running on coal dust ignited by a sparking ignition system, it utilized repurposed steam-engine technology such as cylinders, pistons, connecting rods, and a flywheel. Although prone to overheating and seizing up, Lenoir's invention heralded a power source fundamentally lighter and more compact than steam.

In 1876, German engineer Nikolaus Otto developed the first practical four-stroke engine. Developed with the help of Gottlieb Daimler and Wilhelm Maybach, this engine had four cycles (intake, compression, ignition, and exhaust) and promised higher power-to-weight ratios than earlier internal-combustion engines lacking a compression cycle.

This four-stroke engine of Otto's ran on illuminating gas, the bright-burning fuel of the gaslight era. A stationary power plant, it generated just 3 hp and weighed more than 6,500 pounds (about 3,000 kilograms). Nevertheless, it is the earliest true ancestor of the twentieth-century automobile engine.

Nine years later, Otto contributed a further breakthrough in the form of a low-voltage magneto ignition system for portable engines powered by liquid hydrocarbon fuels. Concurrently, Daimler with Maybach's help pioneered small, lightweight gasoline engines suitable for road vehicles. The stage was thus set for motorized bicycles, motorcycles, automobiles, motorboats, dirigibles, and heavier-than-air flying machines.

Karl Benz, another inventive German, tinkered together the world's first practical automobile in 1885. In addition to building the three-wheeled vehicle and its Otto-cycle engine, he invented almost everything else needed to make it work. The four-wheel Benz Velo of 1894 became history's first volume-produced car. A new industry had been launched and with it a global demand for petroleum.

The emergence of the piston engine struck aerial experimenters as manna from heaven. Here was precisely what they needed: a small and light power source capable of propelling a heavier-than-air vehicle into the sky and keeping it aloft. Unfortunately, this new propulsion technology was so beguiling in and of itself that many aerial dreamers fixated on it and gave short shrift to the rest of flight's challenges.

Those laboring under the chauffeur mind-set envisioned the airplane as a self-righting, self-running vehicle to be driven around the sky according to one's whim. This coming invention would be a gasoline-powered conveyance much like a motor car. How different could they be?

All of this seduced Europe's experimenters in particular, but not exclusively, into believing that flight's primary challenge was getting into the air. Once aloft, the reassuring chauffeur paradigm said, the rest would be easy.

Gabriel Voisin was Europe's premier aerial chauffeur during flight's first decade. Borrowing heavily from Hargrave with Wright influence indiscriminately thrown in for good measure, Voisin constructed gliders and then powered models that swayed and wallowed alarmingly because of insufficient control. They were coaxable more than controllable. Damage was frequent but injuries mercifully few because of the very low speeds and altitudes at which they flew.

In the United States, Samuel Langley likewise fell prey to the chauffeur mind-set and would in fact be its chief exponent in the Americas. Viewing propulsion as the solution to the problem of flight, Langley devoted five years and most of his funding to sponsor development of a lightweight aero engine. By Langley's own calculations, this engine was to weigh no more than 100 pounds (45 kilograms) and generate at least 12 hp.

The only person willing to try to meet Langley's specifications was Stephen M. Balzer, the Hungarian-born mechanical engineer who tinkered together New York City's first automobile in 1894. Balzer created a five-cylinder rotary engine for Langley, but it refused to perform properly. It initially turned out 4 hp and then 8 but not the required 12. With time passing and expenses mounting, Langley's hopes fell.

In stepped Charles Manly, Langley's able assistant, to save the day. A Cornell-trained Virginian in his twenties, Manly converted Balzer's troubled engine from an air-cooled rotary to a liquid-cooled radial. When he finished revising it, the power plant developed 52 hp, an astonishing performance for a turn-of-the-century gasoline engine weighing just 200 pounds (90 kilograms) including its water-filled radiator and associated plumbing.

Now with four times the requested power at his disposal, Langley must have felt success was firmly within grasp. Unfortunately, the airframe he fitted this engine to was largely an afterthought. The Langley Aerodrome A was in fact little more than a fourfold scaling-up of his successful steam-powered models of 1896.

By taking this expedient route, Langley condemned his effort's to failure. Not only did he largely ignore the issue of controllability, but he also overlooked the consequences of scale effects. It should have come as no surprise that his machine failed to fly or that it broke apart the second time Charles Manly tried to fly it. That second plunge into the icy Potomac occurred on December 8, 1903, nine days before the Wrights succeeded at Kitty Hawk.

Compared to the remarkable Balzer-Manly radial, the Wrights' engine was a primitive affair built with the help of mechanic Charlie Taylor, their employee in the bicycle shop. Orville described it thusly:

> The motor used in the first flights at Kitty Hawk, N.C., on December 17, 1903, had [four] horizontal cylinders of 4-inch bore and 4-inch stroke. The ignition was by low-tension magneto with make-and-break spark. The boxes enclosing the intake and exhaust valves had neither water jackets nor radiating fins, so that after a few minutes of running time the valves and valve boxes became red hot. There was no float-feed carburetor. The gasoline was fed to the motor by gravity in a constant stream and was vaporized by running over a large heated surface of the water jacket on the cylinders. Due to the preheating of the air by the water jacket and the red-hot valves and boxes, the air was greatly expanded before entering the cylinders. As a result, in a few minutes' time, the power dropped by more than 75 percent of what it was on cranking the motor.[2]

This engine weighed just under 200 pounds (90 kilograms) and developed 12 hp. The brothers had calculated they needed 8 hp to fly, so this engine actually provided a 50 percent power margin. That they could fly on so little installed power attests to their engineering.

If the Wright 1903 Flyer's engine was a bit crude, the opposite was true of its propellers. The brothers were in fact the first people ever to realize that an airplane's propeller is a rotating wing; as such, it needs to be an efficiently cambered airfoil, not a glorified paddle.

Casting about for available insights to help them create their propellers, they looked to maritime practice only to discover that a ship's screws were designed by trial and error. There being no existing body of scientific theory for them to draw upon, and lacking the time and funding to pursue comprehensive empirical studies, they realized there was only one alternative: they must think it all out, develop their own propeller theory, and then use it to develop their own propellers through calculation.

"What at first seemed a simple problem became more complex the longer we studied it," Orville later wrote. "With the machine moving forward, the air flying backward, the propellers turning sidewise, and nothing standing still, it seemed impossible to find a starting-point from which to trace the various simultaneous reactions."[3]

The brothers did what they always did: they talked it through, arguing points of interest day after day and sometimes well into the evening. Anytime a new thought struck or idea took shape, one brother presented it to the other and a lively discussion ensued. These often ended with each brother having taken the other's position.

After several months of thrashing it all out, the dynamics of propellers were no longer a mystery. "When once a clear understanding had been obtained," explained Orville, "there was no difficulty in designing suitable propellers, with proper diameter, pitch, and area of blade, to meet the requirements of the flyer."[4]

Hand-crafted out of laminated spruce, these two-bladed pusher propellers were 8 feet (2.44 meters) in diameter. They had greater pitch near the hub, where they passed more slowly through the air, gradually flattening out to a mild camber at their tips. The Wrights had scored an aerodynamic bull's-eye, these propellers being within a few percentage points of optimal.

But that wasn't the end to their ingenuity. The Wrights drove these propellers via bicycle chains mounted on sprocket gears. By changing the gear diameters, they could select for a different number of propeller rotations per minute despite an engine that ran at only one speed.

And to balance the thrust, they gave one of the chain drives a half twist so that the propellers rotated in opposite directions.

Whereas chauffeurs such as Langley and Voisin applied power as soon it was available, the Wrights added it last. As true airmen, they taught themselves to fly and achieved control around all three axes before proceeding to powered experiments.

Germany pioneered gasoline engines but then ignored airplanes to concentrate on dirigible engines. Great Britain, a producer of fine automobiles and their engines, likewise showed no interested in aero propulsion. That left the French to single-handedly create Europe's first airplane engines, and they did a fine job of it.

In the summer of 1902, Léon Levavasseur—bearded, heavyset, and pushing forty—took a train from Paris to Normandy. He was a boat builder en route to consult with a friend and possible client. Creative, intelligent, and good with his hands, he had abandoned a beaux arts education years before to pursue his fascination with technology, gasoline engines in particular.

Arriving at Etretat on the Normandy coast, Levavasseur met with Jules Gastambide, the wealthy owner of an electric plant in Algeria

Wilbur Wright's first attempt to fly the 1903 Flyer resulted in damage. Evident in this image are the bicycle-type sprocket chains connecting the engine to the propellers.

who was vacationing at that seaside village with his family. Gastimbide admired Levavasseur because he recognized in him a unique combination of artist and engineer.

Overlooking the English Channel's blue waters, the two friends talked of boats and more as they strolled Etretat's verdant cliffs. Levavasseur expressed his belief that flying machines were possible. What's more, he felt he could develop a lightweight gasoline engine to help make that happen.

Gastambide agreed and offered to fund such a development. It was not as risky as it sounded, since, as Levavasseur pointed out, this new engine could be applied to motorboats as well. Turning to Gastambide's adolescent daughter, he added with Gallic charm that he would name it after her: the Antoinette engine.

Two years later, speedboats powered by new water-cooled Antoinette V-8 engines were sweeping to victory in all of Europe's water races. That same spring, Levavasseur was named engineering director of the first company ever created specifically to produce aero engines. Jules Gastambide was its president and Louis Blériot (soon to depart to continue pursuing his own developments) was vice president.

From mid-1906 onward, 24-hp and 50-hp Antoinette engines powered virtually everything that flew in Europe until Wilbur Wright's arrival two years later. Santos-Dumont's 14-*bis* had a 24-hp Antoinette that was soon replaced by the bigger model. Ferber, Farman, Delagrange, Blériot in his early models, Voisin, Esnault-Pelterie, and others also flew on Antoinette power.

Levavasseur next set his sights on airplane design. Starting in 1907, this artist turned engineer gave the world its first successful monoplane series. Like his engines, they too were called Antoinettes.

Internal-combustion engines are also called piston engines because within each of their cylinders is a piston that goes up and down. In a four-stroke engine with upright cylinders, this piston first moves downward to draw in a mixture of fuel and air. A valve closes and the piston then rises, compressing this mixture within the cylinder. A spark plug ignites the compressed mixture, and the resulting explosion pushes the piston down again. Next a different valve opens and the exhaust gases are expelled as the piston rises, ready to draw in more fuel and air.

The pistons are connected at their base to the engine's crankshaft, which translates their back-and-forth motions into rotational energy. In the simplest airplanes, the crankshaft extends out of the engine to become the airplane's propeller shaft and a propeller is bolted to it. In a more complex installation, the crankshaft connects to a transmission that steps down the engine's rpm so that the propeller turns more slowly for greater efficiency. The propellers can do fancy things too, as we shall see.

As designers created piston engines, they found many ways to arrange the cylinders. If they lined them up one behind the other, the result was an *inline engine.* A six-cylinder engine of this type might be referred to as a *straight six.* Two inline rows, canted inward to turn the same crankshaft, could be mounted side by side on one crankcase to create a *vee engine.*

The cylinders can also be arranged radially around the crankshaft. The cylinder heads of these circular engines fan out like the points of a star. There are two types of engines arranged this way, the *radial* and the *rotary*. Both benefit from having a much shorter crankshaft and crankcase than an inline engine, which reduces the engine's weight. However, these engines' greater frontal area creates proportionately more aerodynamic drag than an inline engine.

The difference between rotary and radial aero engines is striking. Having blazed a trail of glory over war-torn Europe, rotaries all but vanished after the 1918 armistice. In contrast, radials would rise to prominence in the 1920s and play a major role right through to the jet age.

In 1905, Laurent Séguin dropped out of college to join his older half brother Louis in a far more interesting venture. Together they formed an automobile engine company, Société des Moteurs Gnome. Two years later, as aviation interest heightened in France, they set out to develop an aero engine that would be a complete departure from existing European practice.

What these young technologists came up with was a seven-cylinder rotary engine called the Gnome, and it was an odd beast. For starters, what looked like its propeller shaft actually faced backward and bolted to the airplane's firewall. As for the propeller, it bolted to a fixed shaft on the engine's other side. Consequently, when the Gnome ran, its cylinders and propeller spun as a single unit.

Strange as it may seem, there were good reasons for spinning the engine too. One was adequate air cooling, a constant challenge in those early days. Other aero engines required liquid-cooling systems with bulky radiators, but not the Gnome. Another benefit was smooth operation in an era when most engines ran roughly if at all. But the best thing about the Gnome was its high power-to-weight ratio.

First run in the spring of 1909, the Gnome reliably produced 50 hp for a total weight of just 165 pounds (75 kilograms): It was the first successful aviation rotary.[5] Following Gnome's lead, rival company Le Rhône came out with an excellent rotary engine in 1912. Gnome and Le Rhône competed head to head and licensed other firms to build their respective products, among them Bentley, Clerget, Oberursel, and Thulin. Consequently, Italy, Germany, Great Britain, Russia, Sweden, and the United States also produced rotary aero engines.

A Gnome rotary engine spins together with its propeller in this Nieuport 28 as it prepares for takeoff in World War I.

A 1915 merger ended the initial rivalry by creating Société des Moteurs Gnome et Rhône. By then, Europe was embroiled in World War I. Rotary engines played a starring role in this conflict, but for all their benefits, it was their downside that made World War I aviation so colorful. Fuel and oil had to be fed to them through the hollow shaft that was their single point of contact with the airplane. This meant those liquids had to be mixed together, which in turn dictated the use of castor oil, the only high-temperature metal lubricant for which gasoline is not a solvent.

Most engines have an oil sump through which engine oil is recaptured and recirculated. Rotary engines could have no sump, of course, so the castor oil was drawn into the cylinders and burned along with the gasoline. This produced copious quantities of a gummy exhaust that the fast-spinning cylinders sprayed in every direction.

To deal with the mess, rotary-powered World War I fighter planes often had aluminum engine cowlings cut away at the bottom like a horse's collar to direct this exhaust away under the airplane. Even so, pilots invariably inhaled large amounts of castor oil. It being a laxative, diarrhea was a frequent occupational hazard. Fortunately for wartime pilots, blackberry brandy quickly emerged as the preferred antidote.

That filmy exhaust also gummed up windshields and goggles. One reason aviators wore silk scarves was to have something to wipe their goggles with. The other, or so it is said, was to tickle the back of one's neck as a constant reminder to look around for enemy fighters.

Another odd characteristic of rotary engines was that these fast-spinning masses of metal proved to be extremely powerful gyroscopes. The resultant forces made the airplanes veer on takeoff, requiring open fields rather than defined runways. Gyroscopic forces also made it difficult to turn a Nieuport, Sopwith, Hanriot, or other rotary-powered type to the left. On the plus side, these same gyroscopic forces produced astonishingly tight turns to the right.

The short noses of rotary-engine fighters also helped maneuverability by concentrating the airplane's mass near its center of gravity. Together these factors made rotary-powered World War I fighters perhaps the most maneuverable fixed-wing airplanes in history. Their prowess in swirling dogfights was fearsome, although enemy pilots always knew in which direction their adversaries would turn.

Another foible of rotaries is that they lacked throttles and generally ran at full power all the time. To slow sufficiently to land, pilots employed a coupe button or "blip switch" on the stick that momentarily interrupted the ignition. When the button was pressed, the engine fell silent; when released, it erupted into life again because inertia kept it spinning. Care had to be taken not to blip for too long or the spark plugs would foul and the engine would not restart.[6]

A Sopwith Camel returning from a mission would trace a sawtoothed profile as it approached its home aerodrome. It would dip and bob with each successive *brrrpp-brrrpp-brrrpp* as the pilot blipped its engine. Only when sufficiently slowed down would the pilot plunk down the airplane, cut its master switch, and let the tail skid's drag on the grass bring him to a stop.

Wartime urgency accelerated all aspects of flight technology, aero engines included. One way to get more power out of an airplane's engine was to make it turn faster so that there were more explosions in its cylinders per given unit of time.

From 1914 through 1918—the span of World War I—the rotational speed of inline aero engines rose from 1,200 rpm to upward of 2,000 rpm. Unfortunately, rotary engines couldn't pull off this same trick because running the engine faster exposed the spinning cylinders to a self-defeating rise in aerodynamic braking. As the war went on, therefore, inline engines outstripped the capabilities of rotaries. Mercedes manufactured a large number of six-cylinder vee engines for many different German fighter types of World War I. Fine as those engines were, however, the Allies had an even better one, thanks to an inventive Swiss engineer working in Spain.

Born in Geneva in 1878, Marc Birkigt designed mining equipment before relocating to Barcelona at the start of the 20th century. A gifted designer of engines and luxury automobiles, he soon distinguished himself as the engineering genius behind the appropriately named Hispano-Suiza company.

Birkigt opened a factory in Paris in 1911. When World War I broke out a few years later, he turned it over to Gnome for mass production of their rotary engines and returned to Barcelona. It was there that he conceived of an aero engine built a radically different way. The result was the Hispano-Suiza V-8.

Initially developing 150 hp, this engine introduced *monoblock construction*, whereby thin steel liners were screwed into the cylinders of the cast aluminum block, resulting in a strong, lightweight engine that was easy to manufacture. The casting included cored passages for cooling water to circulate. Corrosion was avoided because this water never touched steel. The overhead cams actuating the valves were completely enclosed.

This engine caused a veritable sensation when the French saw it in the summer of 1915. Adopted for military use, Hispano-Suiza engines powered the famous SPAD series, France's most widely produced fighter. The 150-hp "Hisso" powered the SPAD VII and an improved 220-hp version, incorporating reduction gearing, powered the SPAD XIII. A 300-hp version was also developed before the war ended but suffered from teething troubles.

The Hispano-Suiza—the world's first cast-block engine—was the most technologically advanced power plant to emerge from World War I. Built by fourteen firms in France alone, it was also produced under license in Great Britain, where as the Wolseley Viper it powered the S.E.5a fighter, and in the United States.

Birkigt's influence cannot be overstated. Modern automobile engines are the technological descendents of his World War I Hissos. So too were the liquid-cooled aero engines of World War II, including the Rolls-Royce Merlins powering the British Spitfire and U.S. P-51 Mustang as well as the Daimler-Benz engine in Germany's Bf 109.

Lt. John Macready climbed steadily over Dayton, Ohio, home to the U.S. Army Air Service Engineering Division at McCook Field. It was September 28, 1921, and Macready was piloting an open-cockpit LePere biplane.

The test pilot wore several sets of woolen long underwear beneath his regulation Army uniform. On top of that were an electrically heated knit-wool jumpsuit and a down-lined flying suit made of heavy leather. Thick fur-lined gloves protected his hands and fleece-lined outer moccasins his feet. On his head, Macready wore a specially insulated leather flying helmet, face mask, and goggles treated with a gelatin coating to inhibit ice formation. Clenched in his teeth was a pipe-stem mouthpiece attached to a rubber hose. It spewed lifesaving oxygen from a cylinder of compressed air.

The LePere's 400-hp Liberty engine likewise benefited from supplemental oxygen because Macready was testing an exhaust-driven turbo-supercharger. This experimental device permitted the airplane to continue ascending long after the thinning air would have robbed a normally aspirated engine of power.

Macready's altimeter read 41,000 feet (12,497 meters) when his engine sputtered and died, although a recording barograph later showed the true altitude to have been 37,800 feet (11,521 meters). The Army flier glided down to a safe dead-stick landing an hour and forty-seven minutes after taking off.

More than just setting an altitude record, this research flight demonstrated new technology that would prove critical almost two decades later.

The United States developed two aero engines for World War I, both of which were built in great number. The big one was the Liberty, the best version of which had twelve cylinders and turned out 400 hp. It clearly reflected German design influence.

The Liberty engine and the American-built version of Great Britain's de Havilland DH-4 bomber that it powered were the United States' most important industrial contributions to World War I. A half dozen U.S. companies built Liberties, which powered many U.S. aircraft and remained important right up to the 1930s.

The smaller U.S. engine of World War I was the Curtiss OX-5, an obsolescent 1915 design built for the Curtiss JN-4 Jenny military trainer. OX-5s taught many U.S. pilots to fly. Their unreliability also taught them to be constantly on the watch for possible emergency landing fields.

Huge stockpiles of surplus Liberties and OX-5s retarded U.S. aero engine development for much of the 1920s because newer designs and technology could not compete on price. In a dramatic reversal,

Aerial daredevil Gladys Ingle prepares to leap from one Curtiss JN-4 Jenny to another. Flown by barnstormers and aerial circuses during the Roaring Twenties, most Jennys had unreliable OX-5 engines.

however, the air-cooled radial would arise before the decade was out to topple the dominance of liquid-cooled inlines such as the Hispano-Suiza, Liberty, or OX-5.

When Lindbergh conquered the Atlantic in May 1927, he did so with an extraordinary engine that ran flawlessly for thirty-three and a half hours. That remarkable piece of engineering was the Wright J-5 Whirlwind, a nine-cylinder, air-cooled radial power plant weighing 500 pounds (227 kilograms) and developing 220 hp. Introduced just the previous year, the J-5 was history's first fundamentally reliable aero engine. People had achieved reliable airframes a decade earlier; now here was an engine to match, and it existed largely thanks to the U.S. Navy.

In November 1910, a civilian aviator named Eugene Ely took off in a Curtiss Model D from a platform built over the bow of a Navy ship lying at anchor in Hampton Roads, Virginia, landing ashore minutes later. Again flying a Curtiss Pusher in California two months later, Ely flew from shore to land on the specially constructed platform of a modified Navy cruiser in San Francisco Bay.

These experiments demonstrated the potential for airplanes to play a role at sea. Naval aviation was born and began working out the mechanics for shipboard use. After World War I, Congress appropriated funds allowing a coal-carrying vessel to be converted into the USS *Langley*, the first U.S. aircraft carrier. Named after Samuel Langley, the ship was launched in 1922.

In the meantime, naval planners had defined the requirements for the service's aircraft. Because they would go to sea aboard aircraft carriers and space was at a premium aboard those vessels, these airplanes had to be dimensionally compact. And since space for stowing spare parts and performing maintenance would be constrained, they should also be as simple and reliable as possible.

These factors all favored air-cooled radial engines, which dispense with radiators and associated plumbing. That meant lighter weight, less required maintenance, fewer things to go wrong, and fewer spares to keep on hand. Better still, since radial engines are shorter than liquid-cooled engines, the airplanes themselves would also be shorter, allowing more of them to occupy crowded hangars and flight decks.

With this in mind, the Navy got in touch with the Lawrance Aero Engine Company of New York City in February 1920. Lawrance, tiny and virtually unknown, was then the only U.S. firm marketing an air-cooled radial engine.

Charles Lanier Lawrance, born in Massachusetts in 1882, was the scion of an affluent New England family. After attending Yale University, he sailed off to Paris in 1906 to pursue graduate studies in architecture. More than just a world center for art, literature, and music, the French capital was also belle époque Europe's epicenter for new technologies. The excitement must have rubbed off on the young New Englander. Returning home three years later, he abandoned his plans for a career as an architect to pursue a growing fascination with gasoline engines.

By the time the United States entered World War I, Charles Lawrance had developed a number of engines for race cars. Following the war, his interests focused on air-cooled radial engines for the emerging aviation market. There would be demand for the power plants he visualized, and he meant to meet it.

Air-cooled radials already had a small foothold in flight. Louis Blériot had crossed the English Channel in 1909 on one built by Alessandro Anzani, an Italian bicycle racer living in Paris who specialized in lightweight motorcycle engines. Blériot's three-cylinder, 25-hp Anzani ran continuously for thirty-seven minutes, although it threw out copious quantities of oil.

It must be said that no air-cooled radial engine was anywhere near reliable in 1920. That included Lawrance Aero Engine's initial product offering, a forgotten three-cylinder model producing 60 hp, as well as a nine-cylinder experimental radial it had just developed for the Power Plant Section of the U.S. Army Air Service's Engineering Division at McCook Field. The Army too was showing interest in radials.

One can imagine Charles Lawrance's surprise when the Navy came calling for an engine producing 200 hp. Setting to work, he and his team defined another nine-cylinder radial design. The Navy purchased an initial fifty.

It concerned the Navy that Lawrance Aero Engine might not be able to meet its production commitments. While the company had engineering and drafting facilities, it lacked a true factory and instead procured all parts externally. Navy officials approached the nation's two premier engine manufacturers to inquire whether they too would develop air-cooled radials.

Neither the Curtiss Aeroplane and Motor Company nor the Wright Aeronautical Company—founded by the Wright brothers in 1909—was interested in developing a 200-hp radial, which left Lawrance as the Navy's sole supplier. Curtiss and Wright had declined because they were doing well with their current product lines. Curtiss, carrying forward its founder's focus on liquid-cooled vee engines, was on the verge of delivering its large D-12, a 375-hp engine for Army fighters that would produce 500 hp by the mid-1920s. As for Wright Aeronautical, which from 1916 to 1919 was known as Wright-Martin, it had negotiated wartime U.S. manufacturing rights to the Hispano-

Suiza engine and continued to improve on that European propulsion technology.

Wright-Hissos were in strong demand, so there was little incentive for the company to undertake a costly new line of development. Moreover, Wright Aeronautical was then developing an air-cooled radial of its own, the R-1, for Army evaluation at McCook Field. So troubled was that experimental engine that Wright was not eager to take on more.

At this juncture, the Navy did something of seismic consequence that would change aviation for the better. To be certain of sufficient manufacturing muscle behind the Lawrance J-1, it arranged for Wright Aeronautical to acquire Lawrance's firm and build his engine. And with Wright's board of directors dead set against the plan, it required a bit of blackmail.

Naval leaders in Washington, D.C., informed Wright Aeronautical in 1922 that the service would buy no more Wright engines or spares from the firm if it refused to acquire Lawrance Aero Engines. The Navy being a major customer, Wright had no choice but to comply, and in 1923 the J-1 became a Wright product. Charles Lawrance, Wright's new vice president, relocated to its corporate facilities at nearby Paterson, New Jersey.

The handwriting was on the wall. With the Army now also pursuing radial technology, notice had been served on the nation's aero engine industry (then led by Wright, Curtiss, and automotive giant Packard) that the U.S. government would no longer foster two parallel lines of technological development. In the future, air-cooled radials would be favored for public development and procurement funds.

Sam D. Heron spent World War I working at the Royal Aircraft Factory, Britain's flight research center at Farnborough, southwest of London. While there, this Englishman in his mid-twenties participated in the first comprehensive scientific studies ever performed of engine-cylinder air cooling.

Airplanes flew rapidly through the air, so it was logical to want to use the slipstream to cool their engines. Unfortunately, metallurgical knowledge and metal fabrication techniques were not yet sufficiently advanced to allow airplanes—other than those with spinning engines that force-cooled their cylinders—to dispense with the weight and complexity of liquid-cooling systems with radiators.

The son of an actor, Heron had attended night schools but lacked resources for a degree. Instead it was an apprenticeship as a mechanic and foundry worker that won him participation in Britain's wartime developments. While there, Heron helped design the world's first successful air-cooled aluminum cylinders.

Heron was a difficult personality but his genius for engine design was clear. After brief stints at engine makers Rolls-Royce, Napier, and Siddeley, he allowed the United States to hire him away in 1923 as a civilian researcher in the Power Plant Section of the Air Corps' Engineering Division. He found his niche at McCook Field. It was there that he invented the sodium-filled valve, a key technology that made the high-power aero engine possible. Cylinder exhaust ports and valves are the hottest parts of any internal-combustion engine, so it is here that the cooling challenges are greatest. As for the valves themselves, they are disc-shaped metal plugs mounted on stems that rise or fall at high speed to open or block off access to a cylinder as the engine operates. Opened during the exhaust stroke, these valves are exposed to the hot, high-pressure flow of corrosive exhaust gases.

Heron's brilliant idea was to hollow out the exhaust valve (the area behind the valve face and partway back into the stem) and then partially fill this internal void with a liquid substance that would circulate internally as the valve oscillates. This circulation would spare the valve's face by continuously drawing heat away along its stem.

Mercury was the obvious choice for a filling, but it worked poorly. Then Heron found sodium, which is a liquid at engine operating temperatures, and a critical piece was added to the technological puzzle that is the reciprocating engine. Heron would make other contributions to engine design, metallurgy, and fuels development, but this was his most important.

To Frederick B. Rentschler, president of Wright Aeronautical, the Navy's action was a golden opportunity. A talented propulsion engineer with business acumen and an entrepreneurial streak, Rentschler had spent World War I as an Army officer assigned to inspect U.S.-built Hispano-Suizas at the Wright-Martin plant. Subsequently joining the firm as an engineer, he had risen to lead it.

Fred Rentschler had promoted the Wright R-1, the company's first radial project. Rejecting the conventional wisdom that liquid-cooled engines would always be dominant, he saw from the outset that air-cooled radials offered potentially greater simplicity, reliability, power-to-weight ratios, and total horsepower. Unfortunately, he couldn't get his company's conservative board of directors to agree with him. While he was glad now to gain Lawrance's program as well, he thought big and wanted to pursue radials considerably more powerful than the 200-hp J-1.

Wright's board continued to oppose him, so Rentschler quit late in 1924 and took much of the company's engineering talent with him. The following summer they set up shop in Hartford, Connecticut, as the Pratt & Whitney Aircraft Company. Their first product, a nine-cylinder radial called the Wasp, was running before the end of the year.

As for Charles Lawrance, he became Wright Aeronautical's new president. T. E. Smith, former head of the Air Corps Engineering Division's Power Plant Section, arrived soon thereafter to serve as Wright's chief engineer for a couple of years. Smith brought Sam Heron with him, and between them they restored the company's technical expertise.

In short order, America had healthy competition between these two galloping giants of the aviation industry. They turned out successively better versions of this world-changing propulsion technology called the radial aero engine.

The J-1 and its immediate successors had significant teething troubles. Charles Lawrance and his engineering team made constant changes to this evolving product line. The J-4 of 1924—the first to bear the illustrious name Whirlwind—was quite good, although it too suffered from limited reliability.

Then Sam Heron extensively reworked the J-4's cylinder design, creating the J-5 Whirlwind of 1926. Here at last was a radial engine that combined efficient combustion with adequate cooling thanks to its redesigned cylinder head, the use of sodium-filled exhaust valves, and other technical features.

The following year, Lindbergh showed the world just what the Wright J-5 Whirlwind could do (Heron personally visited the *Spirit of St. Louis* and fine-tuned its engine before the young flier's departure). And soon the Wright J-5 Whirlwind was setting records in every corner of the globe.

Wright engineers were already at work on a more powerful radial series called the Cyclone. Pratt & Whitney would soon bring out its Hornet series as well as twin-row versions of its Wasp. These had a second bank of cylinders staggered slightly behind the first for adequate cooling.

A decade later, these two companies' products would power most of the world's airliners and many first-line U.S. military airplanes. By the latter 1930s, in fact, just one liquid-cooled engine would remain in U.S. production, that being the 600-hp Curtiss Conqueror. By then, Curtiss and Wright had joined forces in a 1929 merger to create the Curtiss-Wright Corporation.

Aerodynamic drag was the one area where the radial engine was at a disadvantage compared to the liquid-cooled power plant. The *Spirit of St. Louis* is a perfect example. Sticking out of the monoplane's otherwise sleek nose, this engine's nine cylinders acted like a constant brake in flight. In contrast, airplanes with liquid-cooled engines were aerodynamically clean. Although their engines and cooling systems weighed more and their radiators incurred a drag penalty, they came out ahead in terms of speed per horsepower because they avoided all that cylinder drag.

This relative advantage increased in 1927 when Prestone, the trade name for ethylene glycol, was introduced. With boiling and freezing points respectively much higher and lower than those of water, Prestone allowed for smaller radiators and less coolant, reducing both drag and weight. This advancement allowed the Army's P-1 Hawk pursuit plane of the mid-1920s, which had a deep-bellied front, to evolve by decade's end into the svelte P-6E Hawk.

Just when it looked as if liquid-cooled engines would always have the edge, a young American in rural Virginia made a surprising discovery. A native Chicagoan with an engineering degree from the University of Illinois, Fred Weick worked briefly for the U.S. Navy's Bureau of Aeronautics before accepting a position in 1925 with the National Advisory Committee for Aeronautics.

Established ten years earlier, NACA was fast evolving into the world's premier aeronautical research entity. Because he had worked in propeller design for the Navy, NACA assigned Weick to its Langley Aeronautical Research Center in Hampton, Virginia (today the NASA Langley Research Center).

Now in his mid-twenties, Weick was asked to help design and build a full-scale wind tunnel for testing actual engines and propellers. When it was finished, he became its chief and conducted tests culminating in his writing a seminal book on propeller design.

In 1927, Weick looked at possible engine cowlings for the air-cooled radial engine, which held a lot of promise if their excessive aerodynamic drag could be mitigated. Since airplanes would soon be going faster and aerodynamic drag increases as the square of the airspeed, this was of great concern to the aeronautical engineering community.

For Weick's planned tests, a Wright J-5 Whirlwind engine and propeller were mounted in the Propeller Research Tunnel on a support structure resembling the front of an airplane. NACA fabricators also built seven test cowlings requested by Weick for this study. Addressing the spectrum of possibilities, these ranged from a minimal cowling (a narrow-chord band around the cylinder heads akin to the British Townend ring) to a full cowling encasing the engine.

Before evaluating this spectrum, Weick and his team ran baseline tests of the uncowled engine. The wind tunnel's scales registered 85 pounds (39 kilograms) of drag at an airspeed of 100 mph (160 km/h). This meant that a typical single-engine airplane with an exposed radial expended up to 30 percent of its fuel supply just to overcome engine drag, a patently unacceptable cost for any cooling system. Sobered, the team set to work.

Being methodical, Weick included a full engine cowling of airfoil

In the late 1920s, Fred Weick tested his full NACA cowling in the full-scale propeller research wind tunnel at Hampton, Virginia.

cross section (it was in effect a circular wing) that would enclose the radial engine. Neither he nor anyone else expected this full cowling to provide acceptable cooling. The wind tunnel evaluation showed that while cylinder head temperatures were indeed somewhat higher, the cowling yielded a 60 percent reduction in drag.

If this full cowling could be made to work, it meant an instant reduction in fuel consumption and a concomitant increase in range and operating economy. It also heralded an end to the speed advantage that made liquid-cooled engines the only choice for fighter plane designers. With this new cowling, equivalent airframes with either a liquid-cooled engine or a radial engine of the same horsepower rating would fly equally fast.

Fully cowled radial engines had flown before—for example, the military's experimental Dayton-Wright XPS-1 of 1922. However, nobody before Fred Weick had shown the potential benefits in terms of drag reduction. For this achievement, NACA received the Collier Trophy in 1929.

The next challenge was to achieve acceptable cylinder head temperatures within the aerodynamic NACA cowling. The champion here was Rex Beisel, a talented aeronautical engineer who would later lead the design of the Vought Corsair fighter plane of World War II fame. Heading up a four-year program of wind tunnel and flight research, Beisel learned what in aerodynamic terms was going on *inside* the cowling.

Beisel's solution was to restrict total airflow through the cowling and guide this limited flow via pressure baffling to where cooling was most needed. This effort too was successful; in fact, Beisel stunned the aviation world in 1934 by achieving lower cylinder head temperatures inside a NACA cowling than in the same engine with no cowling at all.

From the 1930s onward, liquid-cooled engines would play second fiddle to radials. This would continue through World War II and all the way to the jet age.

As long as airplanes flew slowly, fixed-pitch propellers were fine. They were made of laminated wood before and during World War I, and increasingly of metal thereafter. As airplanes became faster, however, the growing difference between takeoff and cruising speeds turned fixed-pitch propellers into a frustrating liability.

A fixed-pitch prop is like a bicycle without the ability to shift gears. Depending on what fixed gear ratio exists between the rotation of the pedals and that of the rear wheel, the bicycle can be either easy to start up from a standstill or comfortable to ride along at a fast clip. Unfortunately, though, setting the ratio for the former makes the pedals turn too fast for the latter, and optimizing for the latter makes the pedals too heavy to push at start-up.

Consequently, early bicycling enthusiasts—such as the Wright brothers—rode using a compromise in-between gear ratio determined by the size of the sprocket wheels employed in the bicycle's chain-drive transmission. This one-gear limitation remained until derailleurs were introduced, allowing bicyclists to shift between gears.

It was the same with fixed-pitch propellers. Depending on the pitch angle of the blades, a propeller could be optimized for efficient takeoffs and landings, an efficient cruise, or somewhere in between. This compromise third choice worked well enough until the disparity between the top and bottom ends of the operating speed envelope became too great for one propeller to do it all. Fortunately, help was on the way in the form of Frank Walker Caldwell.

A native Tennessean with a mechanical engineering degree from MIT, Caldwell stumbled upon his life's work in 1912 when he joined the propeller department of the Curtiss Aeroplane Company in Buffalo, New York. Propellers fascinated this young man. Years before World War I, he presciently foresaw the need for them to evolve as airplanes became larger, heavier, and faster. In particular, he was intrigued by the notion of propellers that could change pitch in flight.

So enthralled was Caldwell with this concept that he left Curtiss for government service as a civilian employee with the Army Air Corps' Engineering Division in 1917. Taking up residence at McCook Field, he served as head of the division's Propeller Department. There he pioneered test practices as well as design and manufacturing innovations.

Frank Caldwell was responsible for the aviation industry moving to drop-forged metal propellers built up of separate blades and hubs. Between flights, these ground-adjustable units could be set to a different blade pitch by loosening the hub, rotating the blades to a new common angle, and then securing them again in the hub.

Ground-adjustable propellers let airplane operators set their propellers to favor either takeoff and climb or cross-country cruising. This helped a bit, but the real need was for a propeller that could change its blade pitch while flying. It was an elusive engineering goal, and what Caldwell had seen of early attempts to achieve it mechanically led him to pursue other avenues.

In 1928 he filed patents for a hydraulically actuated propeller. He also left government service for a position with private industry because it alone had the resources and expertise needed to develop, perfect, and market the advanced propeller he envisioned. Caldwell selected the Standard Steel Propeller Company of Pittsburgh, Pennsylvania, which had built the propeller for the *Spirit of St. Louis*. However, he could just as easily have selected its U.S. competitor, Hamilton Aero Manufacturing of Milwaukee, Wisconsin, which was then the world's leading propeller producer.

Ironically, the next year Fred Rentschler of Pratt & Whitney and Bill Boeing in Seattle collaborated to create the United Aircraft & Transport Corporation, America's largest aviation holding company. During this industry consolidation, UA&TC (today United Technologies) acquired Standard Steel Propeller and merged it with Hamilton Aero Manufacturing, which it already owned, to create Hamilton Standard of East Hartford, Connecticut.

The ideas that Frank Caldwell brought to Hamilton Standard (today Hamilton Sundstrand) would make it famous around the world. Caldwell's ultimate goal was to develop a constant-speed propeller that automatically adjusts its blades to any angle for maximum aerodynamic efficiency at any given engine power setting and regime of flight. With a critical need emerging for any workable system, however, Caldwell decided initially to bring out a simpler two-position propeller that could shift between low pitch for takeoffs or landings and high pitch for cruising flight.

Even as the stock market crashed and the Great Depression spread, Caldwell's team at Hamilton Standard completed the world's first variable-pitch propeller. This Hamilton Standard two-position propeller reached the market in 1932. Employing counterweights, it worked reliably in service.

United Air Lines had a problem. Proving trials with its new Boeing 247 in 1933 showed unacceptable field performance at Denver and other high-altitude airports on its young transcontinental route system.

Here was a golden opportunity for Hamilton Standard to show what its new propellers could do. Caldwell led a team to Cheyenne, a town on Wyoming's high plains near the Colorado border. At the Cheyenne airport, some 6,000 feet (1,830 meters) above sea level,

The Boeing 247 was initially delivered with two-bladed, fixed-pitch propellers.

this team took the fixed-pitch propellers off the 550-hp Wasp engines of a Boeing 247 supplied by United and mounted variable-pitch propellers in their place.

Flight tests showed that the new propellers reduced the Boeing 247's takeoff distance by 20 percent, increased its rate of climb by 22 percent, and allowed a more fuel-efficient cruise. Based on these dramatic results, United immediately placed a fleetwide order for the new propellers. So did Douglas for its new DC-2.

A key piece of propulsion technology had been added, but Caldwell was not through revolutionizing aviation. In 1935, Hamilton Standard brought out Caldwell's constant-speed propeller. Compared to the variable-pitch units, which offered the equivalent of a single gear shift in the sky, these new Hydromatic propellers let pilots select whatever rpm yielded optimal aerodynamic performance. The automatic governors of these constant-speed units uniformly increased or decreased blade angle to maintain this selected rpm.

In 1938, Caldwell added *feathering* to the Hydromatic propeller's features. If one of the plane's engines failed, the extreme aerodynamic drag of its stilled propeller increased fuel consumption and impaired the airplane's ability to maintain altitude on the remaining engine or engines. Feathering the blades, or turning them edge-on to the slipstream, greatly reduced this drag penalty, facilitating continued flight. This safety enhancement proved particularly valuable a few years later in World War II.

Another clever individual, a Canadian named W. R. Turnbull, independently invented a variable-pitch propeller in the late 1920s that was electrically rather than hydraulically actuated. Impressed by tests of Turnbull's technology, the Curtiss-Wright Corporation negotiated an exclusive license but did not do anything with it until 1935, when the U.S. Navy ordered Curtiss electric propellers to improve the performance of its flying-boat patrol planes.

Hamilton Standard and Curtiss-Wright kept trying to outdo each other in the constant-speed propeller field. This beneficial competition spurred further innovation and provided a vital technology during World War II.

Remember Army pilot John Macready's 1921 high-altitude flight over Ohio in the supercharger-equipped LePere biplane? Even as the Wright brothers were inventing the airplane, Sanford Moss was at Cornell University building and evaluating what amounted to an embryonic jet engine. This fascination with high-speed gas turbines won him a doctorate from Cornell and a research position with the General Electric Company in its industrial gas turbine division.

During World War I, NACA approached GE and Dr. Moss to develop a turbo-supercharger (today turbocharger) for aviation. By compressing air before feeding it to the engine, a supercharger prevents decreasing atmospheric pressure from robbing an airplane's engine of power as it climbs. Superchargers can either be directly driven by the engine (mechanical supercharging) or—as studied by Moss—indirectly by passing the engine exhaust through a turbine whose extended shaft drives a parallel turbine that pressurizes intake air.

Other countries were also pursuing this idea. In the lead was France, where a clever engineer introduced an aviation turbosupercharger in 1917. Auguste Rateau's design built on decades of experience creating industrial turbines for use in steel mills, mine shafts, and electrical-generation plants. His aviation efforts were not destined to be as influential as those of Sanford Moss only because France did not share America's vision of high-altitude flight.

In 1918, a Liberty engine equipped with Moss' supercharger was carted by cog railway to the top of Pike's Peak in Colorado. It performed flawlessly at an elevation of 14,109 feet (4,300 meters), putting out full rated power despite the thin air. This success paved the way for the flight experiments by Macready and others.

Despite scant interest between the world wars, Moss and GE continued to refine this technology, which finally came into its own in the late 1930s. The GE turbo-supercharger made it possible for the Boeing 307 Stratoliner, history's first pressurized airliner, to cruise above the weather rather than slog through it. It also allowed the Army's B-17 Flying Fortress bombers to operate in the substratosphere, where reduced air resistance yielded true airspeeds above 300 mph (482 km/h).

Air heats up when pressurized, of course, but introducing hot air into an engine's cylinders is not desirable because it robs the engine of power and risks a premature explosion of the fuel-air mixture. Consequently, GE supercharger installations included an intercooler, which is a heat exchanger that cools the supercharged airflow before feeding it to the engines.

GE's turbo-superchargers were a critical technology in World War II. All U.S. heavy bombers were equipped with them, and so were the P-38 Lightning and P-47 Thunderbolt fighters. Also vital to the war effort were engine-driven superchargers, which likewise boosted the power and altitude performance of many Allied and Axis warplanes. All airplanes equipped with Merlin engines (among them the Spitfire, Mosquito, Lancaster, and P-51 Mustang) had engine-driven superchargers. So did most German, Japanese, and Italian high-performance warplanes.

After the war, superchargers contributed to the performance of the final generation of piston-engine airliners. Particularly complex was the Wright R-3350 turbo-compound power plant used in the Douglas DC-7 and late-model Lockheed Constellations. This twin-row,

The Lockheed P-38 Lightning was one of many U.S. World War II airplanes equipped with exhaust-driven turbo-superchargers.

eighteen-cylinder radial had three exhaust-driven turbines that recaptured energy from the exhaust stream and fed it back to the crankshaft for added propulsive power.

By then, aero propulsion had become so excruciatingly complex that keeping it all going had turned into a frustrating exercise in diminishing returns. Propulsion technologies that had reached their zenith in World War II were pushed too far in the 1950s, leading to a precipitous decline in engine and propeller reliability levels.

Fortunately, help was on the way.

The great clipper ships of the latter 1800s were the fastest commercial sailing vessels of all time. Rakish and slender, with five or six courses of sail on high masts, these global greyhounds transported tea, spices, and other high-value cargoes at two or even three times the speed of conventional sailing ships. Flying fully rigged before the trade winds, their speed approached 20 knots (23 mph or 37 km/h).

The United States led in the design and construction of clipper ships, which often returned to New York from China with their precious loads of perishable tea in under a hundred days. Because the season's first cargoes commanded exorbitant prices, there was a great incentive to go fast. However, this capability came at a high cost because clipper ships pushed existing sailing technology to its limit. Carrying so much sail, they were more complex and expensive to operate than other vessels. They needed large crews and worked them very hard. Worse still, they required a great deal of maintenance and, being highly prone to damage, were in constant need of repair. It was with an audible sigh of relief that maritime operators switched over to steel-hulled steamships in the late nineteenth century.

That technological transition parallels the one that commercial aviation undertook as the 1950s came to a close. Airlines too heaved a relieved sigh when switching from piston airliners to jetliners. But just as something grand was lost when tall ships became an anachronism, so too did aviation forever lose a degree of romance with the disappearance of the propeller.

The jet age dawned because a soft-spoken German youth found science more interesting than the military career planned for him. Growing up in Dessau, near Leipzig, Hans Pabst von Ohain successfully convinced his father to let him attend Göttingen University.

Philosophical by nature and manifestly brilliant, young von Ohain completed that world-leading institution's seven-year doctoral program in just four years, receiving a doctorate in physics with minors in aerodynamics, aeromechanics, and mathematics in November 1935.

Long interested in flight, von Ohain took up gliding in school but dropped the sport soon afterward when it was politicized by the National Socialist regime, which came to power in 1933. But that brief taste of gliding set von Ohain thinking, particularly the contrast it struck with a commercial flight he had taken from Köln (Cologne) to Berlin. So clangorously noisy and beset with vibrations was that early airliner that it started him wondering whether there might be a better way to power airplanes.

Dr. Hans von Ohain.

What sprang to mind at Göttingen was the gas turbine engine, a concept then being batted around in academic settings. There were good reasons why such jet engines did not yet exist. One was the difficulty of achieving sustained operation, which demanded a theoretical understanding and deft control of high-speed airflows. Another was the challenge of locally extreme temperatures that would push the limits of existing metallurgical knowledge.

From the lectures of Dr. Ludwig Prandtl, von Ohain knew that airplanes would someday operate at very high speeds. Certain that turbine propulsion would play a starring role in this bright vision of flight's future, the student set about trying to imagine and build such an engine in 1933.

After initially paying a local automotive mechanic to build test devices, von Ohain found unexpected sponsorship through the university. This financial support came from Ernst Heinkel, a German industrialist and bon vivant who had his own aircraft company. After graduating, von Ohain went to work for Heinkel Flugzeugwerke.

In 1937, von Ohain developed a prototype gas turbine that ran on hydrogen. A flightworthy engine soon followed that burned diesel fuel. The Heinkel He 178, a small research prototype, was built expressly for this experimental power plant. Late in August 1939, a week before the start of World War II, Luftwaffe test pilot Erich Warsitz took it aloft to successfully perform history's first jet airplane flight.

In England, meantime, Royal Air Force officer Frank Whittle (later Sir Frank) had been working along parallel lines since the late 1920s. Although he started earlier and secured a landmark patent in 1930, a lack of either government or industry support prevented his ideas from taking wing until May 1941, nearly two years after von Ohain's.

Having worked independently, the two men later became friends and happily shared credit as independent co-inventors of the jet engine. Their engines were remarkably similar overall. Both employed a centrifugal-flow compressor, a scaled-up version of existing technology employed in turbo-superchargers.

A centrifugal-flow compressor is a rotating disc with closely spaced, radially arranged scoops or *turbine buckets.* When this disc spins around the engine's central shaft, these scoops gather air and throw it outward, packing it tight in the surrounding casing. Rounding the compressor disc, this compressed flow angles inward to enter combustion chambers, where it is mixed with fuel and ignites, feeding the engine's continuous combustion.

The result is an exploding efflux that provides the jet's thrust. Before exiting the tailpipe, this hot exhaust passes through turbine blades arrayed around the engine's central shaft, causing the shaft to spin at high speed. This bearing-mounted shaft in turn is what drives the compressor at the front, keeping the cycle going.

Nazi Germany's leaders passed over von Ohain's engine and the He 280 jet fighter that Ernst Heinkel developed from it. Instead they sponsored other lines of development, notably the Messerschmitt Me 262 jet fighter with its Junkers Jumo 004 jet engines.

The Junkers Jumo 004 was the design of Dr. Anselm Franz, an Austrian engineer who ranks among the top tier of turbine-propulsion pioneers. Unlike the Whittle and von Ohain designs, Franz's engine featured an axial-flow compressor. In this alternative design approach, air flows through multiple courses of compressor blades arranged radially around a central shaft instead of being thrown outward around a spinning disc. Consequently, axial-flow jet engines are longer and narrower than centrifugal-flow designs.

Franz's slender engine allowed slimmer, more efficient designs that incurred less frontal-area drag. On the minus side, axial-flow engines are much more ticklish in terms of their internal airflow and are subject to sudden and damaging compressor stalls.

When a parallel BMW axial-flow engine development failed to materialize in time, Franz's Jumo 004 was selected for both the Messerschmitt Me 262 and the Arado Ar 234, history's first jet fighter and jet bomber, respectively. Matériel shortages saw this engine built largely out of sheet metal, which shortened its life in service to somewhere between twelve and twenty-five hours, necessitating frequent engine changes.

The United States did so well in World War II using conventional propulsion technologies that it had little incentive to pursue jets. This is ironic because Sanford Moss' supercharger—in essence, one-third of a jet engine—had already contributed to U.S. leadership in high-temperature metal alloys and turbine fabrication.

Great Britain shared its secret turbine technology with the United States during the war, allowing U.S. construction and evaluation of various jet aircraft. While none was used operationally during the war, the United States displayed an explosive burst of technological development in jet aircraft, jet engines, and high-speed flight beginning in the latter 1940s that left the rest of the world behind.

Putting its supercharger expertise to a new use in the meantime, General Electric improved on the Whittle-type engines it produced under license during World War II. It leveraged that success to join Pratt & Whitney and the world's other traditional aero engine manufacturers in postwar jet engine production. Today, GE is a world leader in this field.

Two factors dictated a transition to turbine propulsion. The first is that jet engines are fundamentally more reliable than piston engines. Instead of back-and-forth stresses, turbine power plants feature smooth, continuous rotation. The second factor was that greater speed was not possible with propellers. It was a matter of fundamental physics. World War II fighters had flown so fast that, although they remained subsonic, the tips of their propellers encountered transonic or even supersonic flow. When this happened, shock waves formed. The airflow separated from the blades and the resultant turbulence turned high thrust into high drag.

Postwar propeller airliners were somewhat slower than World War II fighters, but they too ran into trouble. The reason was that their designers were after payload and range as well as speed. This dictated high takeoff weights, which in turn called for powerful engines and propellers capable of accepting that raw power.

One way to get a propeller to handle more power was to increase its diameter. However, that drove up the tip speeds, triggering efficiency losses at lower cruising speeds. Going to larger-diameter props also complicated airframe design by requiring greater engine spacing on the wing and longer, heavier landing gears for adequate ground clearance.

To avoid those complications, designers could instead give a propeller more blades to handle the added power. This kept the propeller diameter the same, but the blades ended up too close to each other. The result was lost efficiency because, denied sufficient time and forward motion, each blade encountered the affected airflow of the previous one.

In desperation, designers resorted to mounting two propellers, one immediately behind the other, on concentric and counterrotating shafts. Very late-model Spitfires equipped with Rolls-Royce Griffon engines show this practice. It worked aerodynamically, but the excessive complexity, high maintenance, and poor overall reliability of these installations made them more trouble than they were worth.

As the 1950s drew to a close, it was obvious that propellers had carried first-line military and commercial aviation about as far as it could go. Higher speeds and gross weights required jet propulsion.

The Boeing 707 debuted in transatlantic service on October 26, 1958. An instant sensation, it revolutionized air travel and launched a rapid conversion to jets. In its first years of operation, however, the 707 assaulted eardrums, rattled windows, and smeared the sky with dark sooty trails. This was true of the Douglas DC-8 and all other first-generation jetliners as well. The reason was their *turbojet* engines, which were earsplittingly loud even fitted with noise suppressors. Those "straight turbojets" accelerated a small amount of air to very high velocity, which is inefficient at low speeds (it was like trying to start a car in high gear). Worse still, their hot exhaust produced a flaring blowtorch noise on contact with the surrounding cold air.

In 1960, a better idea took flight: the *turbofan*. Turbojets and turbofans have the same three internal sections: compressor, combustion chamber, and turbine or "hot section." To this, however, the turbofan engine—also called a *fanjet*—adds a fan unit at front that is larger in diameter than the rest of the engine. Driven by an extended shaft, this bladed unit accelerates a larger volume of air more slowly for improved acceleration and fuel efficiency.

Fanjets are quieter than turbojets because some of the fan's air is ducted around the engine's core rather than through it. This provides an insulating blanket of bypass air between the hot exhaust and cold ambient air. As for the sooty trails, improvements in combustor technology quickly eliminated those.

The first fanjet engines were low-bypass-ratio turbofans that ducted a relatively small amount of air around the core. Over the decades since then, engine developers have moved to higher bypass ratios and incorporated a spectrum of other technological enhancements. The result has been truly dramatic improvements in terms of fuel efficiency, airplane range, and noise.

There are some surprising things to know about modern fanjets. One is that propulsion reliability in the jet age does not depend on engine size, as it did in the piston era, when the most powerful radial engines were considerably less reliable than their lower-horsepower cousins. Modern aero engines are astonishingly reliable. The Boeing 777 world fleet offers a good example. In service since the mid-1990s, there are more than seven hundred of these twin-engine commercial transports in service as of this writing. These 777s are powered by Rolls-Royce, Pratt & Whitney, or GE fanjet engines. The GE90 fanjets powering the most capable 777 models are the most powerful turbine engines yet built. Each generates up to 115,000 pounds of static thrust.

Regardless of which engine type is fitted to its wings, a 777 will on average log about 25,000 flights and spend well over 100,000 hours aloft between each in-flight engine failure or precautionary shutdown. Most 777s will not reach these thresholds before being retired from service.

As for airline pilots, over a full career they might typically amass up to 25,000 logged flight-hours. Even if reliability rates stop improving, most pilots just starting out today can reasonably expect to fly an

The Boeing 777, the world's largest twin-engine airplane, has enormous fanjet engines that produce up to 115,000 pounds of thrust each.

entire career without ever experiencing an in-flight engine shutdown. Nevertheless, crews of course train for them.

Even if a shutdown does occur, the remaining engine can take over without strain. This is ensured by rigorous government certification requirements that dictate twin-engine commercial transports must be able to suffer an engine failure at the worst possible time—on takeoff, when the airplane is flying slowly and at its heaviest—and climb out safely. For this reason, all twin-engine jetliners are 100 percent overpowered.

The planetary emergency that is global warming today focuses increasing attention on jetliner emissions. While great strides have been made over the years to increase fuel efficiency and reduce emissions of carbon dioxide and oxides of nitrogen, these continuing efforts are undercut by ongoing growth in the size of the world commercial fleet.

Aircraft and engine manufacturers, airlines, government agencies, and nongovernmental organizations are today collaborating to address the challenges of global warming. One area being examined is the substitution of more benign biofuels for kerosene, aviation's current fuel. Also promising is the technology now being introduced by the ultraefficient Boeing 787 Dreamliner, the state of the art in aeronautical expertise. The 787 will use about 20 percent less fuel, and create correspondingly lower emissions, than similarly sized airliners. How it does this is the subject of a later chapter.

11 LANDING GEAR

SHOES, CANOES, AND CARRIAGE WHEELS

And once in a while when his landings are rusty,
I always come through with, "By gosh, it's gusty!"

—"THE COPILOT," A 1941 POEM BY DC-3 COPILOT KEITH MURRAY

William Henson was the first human being ever to sit down and try to design an actual airplane. However unrealistic that might have been way back in 1840, it also made him the first person to wrestle with practical considerations such as takeoffs and landings. He gave his Aerial Steam Carriage a set of wagon wheels, that being the obvious choice for a flying stage coach. Tension-spoke wheels would have been better from a weight standpoint, but George Cayley wouldn't get around to inventing those for another decade or so.

Henson used just three carriage wheels in a wheelbarrow configuration to keep weight low. In this regard at least he was far ahead of his time, for he had just given aviation the *tricycle landing gear*.

Nearly all airplanes built today feature the tricycle landing gear. This configuration places the main wheels behind the airplane's center of gravity and has a smaller wheel or wheels at front. Aside from a few crop dusters, homebuilts, and sport airplanes, this configuration is all but universal.

That was not always the case. Aviation initially standardized with the main wheels forward of the center of gravity and a skid at rear (later replaced by a tailwheel). Known as the *conventional landing gear*, that configuration predominated before World War II. There were exceptions then as now, of course, but in general "taildraggers" ruled before World War II and "tri gears" after the war. As for World War II itself, one finds examples of both, although the former predominated.

The conventional landing gear has a drawback. Since the weight is behind the main wheels, the airplane must be perfectly lined up fore and aft with the runway and the pilot must have killed any lateral drift before touching down. If not, the center of mass will try to swing around in front of the wheels, causing the airplane to veer off the runway in a humiliating and potentially damaging horizontal circle known as the ground loop.

Early pilots didn't worry too much about ground looping because airports back then were open fields. Consequently, takeoffs and landings could always be made directly into the wind, keeping things simple. Only later, as the world moved to defined runways, did pilots have to hone their crosswind landing techniques.

There is genuine satisfaction to flying taildraggers. It demands skills and offers joys not found elsewhere in aviation. Landings in gusting or quartering crosswinds are particularly challenging, requiring pilots to be alert and quick on the controls, the rudder pedals in particular. Since side gusts on the vertical tail turn the airplane like a weathervane, it is often said that tailwheel landings are not over until the airplane is parked and tied down.

In contrast, if tricycle-gear airplanes are landed not properly lined up, the center of gravity is forward of the main gear, so momentum wrenches the wheels around to make the airplane track straight ahead. Another benefit of tricycle gears is that the airplane sits level when on the ground instead of nose high. Not since the days of the DC-3 have boarding airline passengers had to hike uphill to claim their seats.

Otto Lilienthal needed no wheels. Like all hang-glider pilots, he took off and landed on his own two feet. Stout brogans were his landing gear of choice.

In the air, Germany's birdman resembled a bird of prey. The reason is his dangling legs, which he threw around to shift his direction of flight. Because birds tuck their legs back in flight for reduced drag, those flailing legs evoked extended talons rather than normal flight.

It was a different story when the Wright brothers built and flew their first two gliders at Kitty Hawk. Not needing to shift their weight for control in the air, they tucked their legs up in flight like a bird. In turn, European experimenters put the Wrights to shame by adopting wheeled undercarriages from the outset. The Wrights stuck with skids far too long, perhaps because they viewed their airplanes as scientific proof-of-concept vehicles first and practical machines second.

In 1609, English explorer Henry Hudson sailed up a North American river that would bear his name. Three centuries later, New York City—located at the mouth of the Hudson River—staged a transportation-themed celebration to mark both this anniversary and steamboat inventor Robert Fulton's navigation of the same waters in 1807.

Held in the fall of 1909, the Hudson-Fulton Celebration treated New Yorkers to the astonishing sight of Wilbur Wright circling the Statue of Liberty. Conspicuously visible between the skids of his Wright Model A biplane was a canoe. Although placed there just for buoyancy in the event of a forced landing in the water, that canoe brought to mind the possibility of boat-hulled airplanes capable of operating off water.

Wilbur flew around the Statue of Liberty on September 29, 1909. On October 4, he flew up the Hudson to a point north of Grant's Tomb and returned to land on the parade ground at Governor's Island. He had covered 21 miles (34 kilometers) in thirty-four minutes. It was only the second time in history and the first in the Americas that an airplane had flown any distance over water.

Two months had elapsed since Louis Blériot conquered the English Channel to win the £1,000 prize posted by London's *Daily Mail* newspaper. Blériot had made that flight at first light to avoid any wind. The year before, *Daily Mail* publisher Lord Northcliffe had contacted Wilbur in France to urge him to compete for the prize, even going so far as to offer considerably more money than the stated prize. Although tempted, Wilbur declined based on the unreliability of the era's engines, and Hubert Latham's subsequent ditchings suggested it may have been a wise decision. The Wrights also passed up requests to participate at Reims in the summer of 1909, although two of their airplanes flew in the hands of others.

Having won the Reims air meet's crowning Gordon Bennett Cup race, a triumphant Glenn Curtiss also participated in New York's Hudson-Fulton Celebration. Like Blériot, though, Curtiss lacked an airplane capable of operating in strong or gusty wind conditions. On September 2, 1909, Curtiss took off from Governor's Island to duplicate Wright's circling of the Statue of Liberty. Heading across the choppy gray waters of New York Harbor, Curtiss found himself blown so wildly about that he abandoned the attempt and landed visibly shaken.

Wilbur's long Hudson River flight two days later took place in conditions so rough that observers saw his machine tossed "upwards and downwards like a ship in a gale," as one reporter put it.[1] Unfazed, the elder Wright brother prepared to fly again that same day. But as he and mechanic Charlie Taylor propped the canoe-equipped Flyer, an explosion blew a cylinder off the engine. It shot straight upward, punching a hole through the top wing, and landed a few paces from Wilbur's feet.

The Flyer's participation in New York's celebration had come to an abrupt end. It didn't matter. Wilbur's courageous display of airmanship had shown the world what only he and Orville could do. Hailed by all America, it was the last public display by the Wrights and a triumphant note on which to exit the world stage.

As for Curtiss, he returned the following year and in better conditions flew the Hudson from Albany to New York City. This record flight covered 150 miles (240 kilometers), during which Curtiss had more time than he might have liked to contemplate the water below.

Glenn Curtiss saw great opportunity for the first person to perfect an airplane capable of operating off of the water. Able to dispense with prepared flying fields, such a machine would also be safer in the event of an engine failure along the coast or over rugged terrain. Given lakes and rivers, both forested countryside and communities without airports could have air service.

Unlike Wilbur Wright, who carried a canoe only for flotation in the event of an emergency, Curtiss decided to see whether an airplane could actually take off and alight on water that way. The resulting machine was so heavy that it failed as a boat and airplane alike. Curtiss then experimented with pontoons that he fitted to his *June Bug*. Thus equipped and renamed the *Loon*, it failed to rise from the waters of Keuka Lake at Hammondsport.

Deciding that warmer water and relief from upstate New York's harsh winters would be beneficial, Curtiss relocated in 1910 to North Island near San Diego, where he had employees training U.S. Navy pilots to fly Curtiss Pushers. It was in San Diego the following year that Curtiss successfully took off from and alighted on water, a first for the Americas. Almost a year had passed since Henri Fabre accomplished that same feat near Marseilles, France.

The Navy took an immediate interest in Curtiss' hydroaeroplane. He soon added wheels to its central float. These allowed the pilot to

The Curtiss A-1 Triad of 1911, the first U.S. aircraft with a retractable landing gear, had wheels allowing it to taxi up onto land or back into the water.

taxi into the water or back on to shore again. Purchased by the Navy as its first airplane, this first amphibious machine was the Curtiss A-1 Triad of early 1911 (the *tri-* in its name denoted land, water, and air). Japan, Germany, Russia, and Great Britain also purchased Curtiss Triads, which started many nations on the path to naval aviation.

Curtiss soon came up with two other water-related innovations. The first was his idea for a stepped float, which is a pontoon whose underside features a tapered cutaway near the back. As the airplane gathers speed, this float rises in the water until the step is exposed, breaking the water's suction. This allows the airplane to skim freely on top of the water and climb out more easily.

Curtiss' second innovation was the boat-hulled fuselage. Perfected back at Hammondsport, this concept had the fuselage itself sit in the water on a watertight hull contoured like Curtiss' stepped pontoon. Before 1912 ended, this idea had evolved into the Curtiss F-Boat, history's first flying boat. The U.S. Navy, Army, and wealthy private owners all purchased F-Boats.

Other designers copied Curtiss' idea. One was Thomas Benoist, whose company in St. Louis provided flying boats for the first scheduled commercial air service in the world. The St. Petersburg–Tampa

The Curtiss F-Boat of 1912 was history's first boat-hulled airplane.

Air Boat Line in Florida began operations on January 1, 1914, carrying one passenger at a time across Tampa Bay in its two-seat, open-cockpit Benoist flying boats. More than twelve hundred passengers were carried in perfect safety before the airline ceased operations that May.

Another was Grover Loening, who earned America's first-ever aeronautical engineering degree from Columbia University in New York. During the 1920s, Loening built rugged biplanes for civil and military customers. The planes' boat-hulled metal fuselages featured projecting prows beneath their propellers. Reflecting his expectation of its use, he called this series the Loening Air Yacht.

In 1914, the Curtiss Aeroplane Company built a twin-engine biplane flying boat for department-store heir Rodman Wanamaker, who envisioned a flight across the Atlantic in the spirit of peace. The *America* was co-designed for him by Curtiss and Englishman John Cyril Porte, a seaplane pioneer in the Royal Navy. Lacking range, this airplane would have depended on strategically placed ships for refueling during landings at sea. Ironically, World War I preempted the noble if precarious attempt.

Curtiss manufactured large flying boats for the Navy, which were the only U.S.-designed airplanes to participate in that conflict. England built Curtiss flying boats under license with major improvements by Porte, launching successful British lines of flying-boat development.

In the months after World War I, the six-man crew of the U.S. Navy's Curtiss NC-4 flying boat conquered the Atlantic by air for the first time. With a wingspan of 126 feet (38 meters) and a gross weight of 21,500 pounds (9,772 kilograms), and powered by four Liberty engines, the NC-4 was a sizable machine. It and two sister ships left New York on May 8, but mishaps prevented the others from completing the voyage. Surviving damage, fog, and other hazards, the NC-4 arrived in Plymouth, England, via Nova Scotia, Newfoundland, and the Azores on the last day of the month.

Many nations would build a great variety of military and civil flying boats in the 1920s and 1930s. They tended to be larger machines than landplanes because they weren't constrained by airfield lengths.

But with engine development chronically lagging behind that of airframes, some of these flying boats turned out to be *too* big.

The Dornier Do X was the ultimate scaling up of metal-airplane pioneer Claudius Dornier's flying-boat design formula. First flown in 1929, this giant transport was 131 feet (40 meters) long, had a 157-foot (48-meter) wingspan, and weighed a maximum of 105,000 pounds (48,000 kilograms). Power was supplied by twelve engines mounted front to back in six strut-supported nacelles atop the wing.

Tested on the Bodensee (Lake Constance), where Germany, Switzerland, and Austria meet, the Do X lifted 169 people in October 1929. Despite this feat, it was underpowered and performed so poorly that it could only cruise at low altitude. Luft Hansa (today Lufthansa) had no interest in the Do X, which likewise failed to find orders on a tour through the Americas at the start of the 1930s.

Advancing technology made this next decade the heyday of boat-hulled airplanes. Pan American Airways conquered first the Pacific with island-hopping services and then the Atlantic, using four-engine *flying clipper ships* built by Sikorsky, Martin, and Boeing. These audacious services pushed the era's airplane range capabilities to the maximum.

Too heavy, the Dornier Do X of 1929 performed poorly despite twelve engines.

Other flying boats also transported passengers before World War II, notably the Empire series developed by British manufacturer Short Brothers (today based in Northern Ireland). Created to link Britain with Africa, India, Australia, Singapore, and Hong Kong, these deep-bellied transports flew littoral routes, which provided opportunities for landings almost anywhere.

Short, Sikorsky, and most other flying boats featured floats mounted on the wings to keep them on an even keel in the water. In contrast, Pan Am's Martin M-130s and Boeing 314s used Claudius Dornier's excellent idea of *sponsons*, which are stub wings set low to also serve as floats, eliminating the need for drag-inducing wing floats.

The crowning culmination of the flying-boat era was the Boeing 314 Clipper, which entered service in 1939. Here was the finest and most capable flying boat of them all, a plane that could fly 3,600 miles (5,800 kilometers) at a stretch. In contrast, the Short Empire series flew 1,500 miles (2,414 kilometers) and could cross the Atlantic only when filled with so much fuel that no payload could be transported.

In an effort to fly airmail between England and North America on the eve of World War II, the British took to mounting a four-engine floatplane atop an even bigger four-engine flying boat, these being the Short Mercury and Short Maia, respectively. Collectively they were known as the Short Mayo composite aircraft. Imperial Airways used this cumbersome but successful system beginning in 1938. The Mayo would lift off from Foynes (today Shannon) on Ireland's west coast. Once aloft and at altitude, the smaller Mercury would start its engines and—its tanks not depleted by the takeoff or climb—head west for a water landing at Montreal, Canada, some twenty hours later.

The Germans, who had inaugerated the world's first aerial passenger services across the North and South Atlantic with dirigibles, had already pioneered a different kind of hybrid airmail service. In 1929, they began catapulting floatplanes off the decks of the *Bremen* and *Europa* when those liners were still far from their destinations. It shaved two days off the arrival of international mail.

Flying boats flew with many nations' military services during and between the world wars. Growing in size and capability, these winged ships performed maritime patrol, antisubmarine warfare, air-sea rescue, and other duties. During World War II in particular, Germany, Great Britain, Japan, the United States, and other nations flew a wide variety of boat-hulled airplanes in all parts of the world, from the Arctic Circle to South Seas atolls.

Glenn Curtiss' concept of retractable wheels for water *and* land operations by the same airplane was realized most fully by the boat-hulled Consolidated PBY-5 Catalina, whose retractable wheels also allowed it to use runways. All Catalinas had cleverly designed wing-mounted floats that folded up in flight to become low-drag wingtips.

For all its use of flying boats, World War II also spelled an end to this technology by spurring the development of long-range land planes and fostering the construction of airfields around the world. It was just as well; as airplane speeds and ranges grew, the weight and aerodynamic penalties of boat hulls became increasingly difficult to justify.

A flurry of excitement attended Howard Hughes' unveiling of his one-of-a-kind Hughes H-4 Hercules flying boat in 1947. With eight engines on wings stretching 320 feet (98 meters), it was by far the largest flying boat ever. To this day, no other airplane has had as great a wingspan. Nicknamed the *Spruce Goose*, this wooden transport flew just once in a brief straight-ahead hop in November 1947.

There were some success stories in the postwar era. Grumman built the Mallard for civil markets and the excellent Albatross for the U.S. Air Force, Navy, and Coast Guard. In Japan, Shin Meiwa developed a turboprop flying boat, and an amphibious version of the same craft, that temporarily gave the concept a new lease on life.

The boat hull even flew in the jet age, starting with Saunders-Roe's single-seat fighter of 1947 and culminating with the Martin P6M SeaMaster, a mid-1950s jet bomber with swept wings and four turbojet engines. A dozen SeaMasters were built for the U.S. Navy. Resting in the water on deep, narrow hulls supported by anhedral wings tipped with floats, they could exceed 600 mph (about 1,000 km/h) in flight and cruise at 40,000 feet (12,192 meters).

In 1876, samurai stopped carrying swords in Japan, Mark Twain published *Tom Sawyer* in America, and the retractable undercarriage was invented in France.

Alphonse Pénaud, confined to a wheelchair, dreamed of a coming invention that would be the very definition of freedom: the airplane. With the help of his friend, mechanic Paul Gaucho, he drew up plans for an astonishingly advanced full-scale flying machine. As described earlier and detailed in their 1876 patent, the Pénaud-Gauchot airplane looked like something out of a science-fiction movie. A two-seat amphibious flying wing, this craft had an enclosed cockpit complete with instruments and flight controls. It had a moth-like wing form, a rudder and elevators, and four-bladed propellers driven by an internal-combustion engine, something only then being invented.

The design also featured fully retractable wheels. Pénaud probably made them that way for the same reason that Glenn Curtiss gave the Triad retractable wheels more than three decades later: so that the airplane could fly from water or land. But even if Pénaud didn't propose this idea for aerodynamic reasons (and he may well have), his patent nevertheless remains the first documented reference to retractable wheels for a heavier-than-air vehicle.

The first airplane to even have wheels in North America was the AEA *White Wing* of 1908. Designed by Frederick Baldwin, a young member of Alexander Graham Bell's Aerial Experiment Association, the *White Wing* flew several times before being wrecked. Its longest flight covered 1,017 feet (310 meters) with fellow AEA member Glenn Curtiss at the controls.

Curtiss had already taken the lead in designing the AEA's next airplane, his *June Bug*. Three years later, he would introduce the Curtiss A-1 Triad with the first wheel-retraction system in U.S. aviation. That same year, the Wiencziers monoplane racer in Germany reportedly also incorporated an experimental form of gear retraction.

Adventurer James Vernon Martin had run off to sea in his teens only to return and enter Harvard University in 1908. In his mid-twenties, suffering no shortage of ego, he called himself "Captain Martin."

Aviation was everywhere in the press that year thanks to Wilbur Wright's performances in France. Completely seduced, J. V. Martin founded the Harvard Aeronautical Society and organized the Harvard-Boston Aviation Meet of 1910. Learning to fly in England soon thereafter, he returned to barnstorm in the United States and start a small aviation firm.

As America entered World War I, Martin talked his way into a government contract with the U.S. Army Engineering Division at McCook Field. A natural promoter, he pressed his idea for a miniature fighter plane capable of intercepting high-flying dirigibles. Although it was something America did not need, the Army provided experimental funding to see what he would come up with.

The result was the J. V. Martin K.III Kitten, a tiny single-seater weighing 350 pounds (160 kilograms) empty and powered by a 45-hp engine. Equipped with oxygen cylinders and provision for electrically heated flight clothing, it had several novel features that raised questions about Martin's competence as a designer.

The first was a structurally questionable wing-strut configuration, and the second was pivoting wingtips instead of ailerons. But it was this little machine's third distinction that would bring it into the record books, because the Kitten also featured retractable main wheels.

Mechanically actuated by a hand crank in the cockpit, this landing gear rotated up and aft to tuck the wheels into form-fitting covers along the fuselage sides. Even with the gear retracted, the bottom half of each wheel protruded into the slipstream. This arrangement allowed emergency wheels-up landings with little resultant damage.

The Douglas DC-3, Boeing B-17, Beech 18, and many other airplanes have since shared this arrangement, which reduces damage when bellying in. In some cases, it also afforded some measure of steering through individual use of the retracted wheels' brakes.

Martin's pride in his landing gear is evidenced by the form-fitting fairings he carefully affixed behind the wheels where they protruded into the slipstream while retracted. This aerodynamic refinement looks oddly out of place on the K.III's otherwise crude airframe.

Completed too late for World War I, the Kitten was tested in the summer of 1919 and proved capable only of brief, ground-level hops of under 300 feet (90 meters). Even so, it remains the first U.S. land airplane to feature retractable wheels. Martin received a U.S. patent for his landing gear in 1916.[2] Today the J. V. Martin K.III Kitten is preserved in the collection of the National Air and Space Museum, Smithsonian Institution.

In 1922, the Verville-Sperry R-3 and Dayton-Wright XPS-1 both flew with retractable wheels. Like the Kitten, they were experimental U.S. Army concepts. Despite this early interest, however, retractable wheels did not come into general use until the early 1930s because before then airplanes flew too slowly to justify their added weight, complexity, and maintenance requirements.

The world's first all-metal, semi-monocoque production airplane was of course the Northrop Alpha of 1930. It seems odd that this bold airplane, which cruised at 145 mph (233 km/h), lacked retractable wheels, but Jack Northrop was reluctant to cut into his multicellular wing for fear of compromising its strength. To mitigate drag, most production Alphas at least had their fixed gear legs encased in aerodynamic fairings.

The Boeing 200 Monomail, which also flew in 1930, was similar overall but did have retractable wheels. Boeing also applied retractable gears to its Boeing 247 airline and YB-9 bomber immediately thereafter. Douglas, Martin, and other companies followed suit with their commercial and military offerings.

Almost from the outset, therefore, retractable wheels were part of the revolutionary formula that was semi-monocoque design. However, the value of this innovation was proportional to aircraft speed, not method of construction, and biplanes too were flying quite fast by the early 1930s. A good example is the Grumman FF-1, which entered fleet service in 1933. The Navy's first fighter with retractable wheels, the FF-1 had a top speed of 207 mph (333 km/h).

The first U.S. commercial airliner with retractable wheels was in fact a biplane. Equipped with powerful radial engines and billed as a high-speed transport, this was the fabric-covered Curtiss T-32 Condor II. As described earlier, the Condor entered service in 1933 only to be rendered hopelessly obsolete by the Boeing 247 later that year and the Douglas DC-2 the year after that.

The Laird Super-Solution racer in which Jimmy Doolittle won the

Bendix Trophy transcontinental speed dash of 1931 had retractable wheels. Flying this green and yellow biplane, Doolittle became the first human being ever to cross the United States in less than twelve hours (seven years earlier, he'd been first across in less than twenty-four hours in a DH-4 biplane). Doolittle's 1931 triumph—a featured event of that year's National Air Races—highlighted the value of retractable wheels in the public mind.

Designers turned to retracting the tailwheel or nosewheel, not just the main landing gear, as aerodynamics became more important. In World War II, they also adopted flush riveting for high-performance fighters, a technology also applied to commercial and military transports as their performance increased with the advent of swept wings and turbine propulsion.

Landing gears both drive the design of the airplane and are driven by it. An example of the former is the configuration of a commercial jetliner, which features a low wing in part because it provides a good place to house the landing gear. Gear legs being moment arms, a gear twice as long must be more than twice as robust in order to resist leverage forces. Low wings thus ensure shorter, lighter landing gears.

As for the latter, military cargo jets provide a good example. They must employ a high-wing configuration to keep the fuselage low to the ground for easy loading. This dictates a sharply upswept rear fuselage to avoid striking the runway during takeoff or landing (a particular hazard for low-slung airplanes). It also accommodates rear doors and a boarding ramp for vehicles and cargo.

To avoid having an excessively long landing gear because of the high wing, designers often house the military transport's main wheels (and sometimes additional fuel as well) in bulged blisters along the sides of the fuselage. This accommodation saves considerable weight and airframe space at the cost of a slight increase in frontal-area drag.

An even more extreme example of how design requirements can dictate landing gear decisions is the Boeing B-47 Stratojet, which flew at the end of 1947. This revolutionary jet bomber's swept wings were too thin to house either fuel or landing gear. Consequently, Boeing gave the B-47 two sets of retractable main wheels housed in tandem in the fuselage, and small outrigger wheels that extended from the inboard engine nacelles to prevent the airplane from tipping laterally.

Aeronautical engineers who work in landing gear design contend with many challenges. One is materials, since landing gears are subject to high stresses and must be enormously robust. Another is safety and reliability. Yet another is runway loading, which is a measure of how much weight each given area of tire contact imposes on an airport's runways, taxiways, and flight ramps. As airplanes grow in size, designers must spread their weight over more tires to keep this loading low.

The Airbus A380 provides a good example. This superjumbo, the largest jetliner in commercial service, rides on twenty-two wheels versus eighteen for the Boeing 747 and fourteen for the Boeing 777. It has four main gear posts, two under the fuselage with six wheels each and two more under the wings with four wheels each. The wing gear posts are positioned slightly forward of the fuselage gear posts. As is universally the case with jetliners, the A380's nose strut has two tires and pivots for steering. Small deflections follow rudder-pedal inputs on the ground, while more significant turns are commanded by a nosewheel tiller in the flight deck. The rearmost axle of each of the A380's fuselage landing gears also deflects to help this giant airliner turn.

Bigger still is the Antonov An-225 Mriya, a Russian strategic airlifter originally developed to transport oversize loads for that nation's space program. Today in commercial service, the An-225 features two rows of seven wheels each on each side of the fuselage. Two side-by-side nose landing gear struts each mount another two wheels, for a grand total of thirty-two. This arrangement is necessary because by virtually any measure the An-225 is the world's largest airplane. This aerial behemoth has a maximum gross weight of 1.4 million pounds (635,000 kilograms), a wingspan of 290 feet (88 meters), and an internal cargo capacity of 550,000 pounds (250,000 kilograms).

12 PASSENGER CABIN

VOYAGING ALOFT

. . . That I wouldn't be surprised to see a railroad in the air,

Or a Yankee in a flyin' ship a goin' most anywhere.

—"THE OLD WAY AND THE NEW,"
A POEM BY JOHN H. YATES (1837–1900)

On January 1, 1914, the world's first airline began operations in Florida. The St. Petersburg–Tampa Air Boat Line was a tiny outfit that transported one paying passenger at a time across Tampa Bay in a Benoist Model XIV. The one-way fare was $5.

The Benoist combined a wooden boat hull with biplane wings, a brass radiator, and a 75-hp inline engine turning a pusher propeller. The clattering engine was immediately behind the seats, and colorful pennants festooned the wing struts. The single passenger stepped aboard, helped by the proto-airline's pilot, Tony Jannus, a twenty-four-year-old with a ready smile. Settling beside him in the open cockpit, this early air traveler found himself or herself exposed to the elements without so much as a windshield for protection.

It was thrilling, if not nerve-racking, but Jannus' reassuring manner quickly put people at ease. Like the driver of a car, he leaned

Carrying one passenger at a time, Florida's St. Petersburg–Tampa Air Boat Line inaugurated the world's first passenger service in January 1914.

his elbow casually over the air boat's gunwale as he taxied out onto the bay. Passengers could drag their hands in the water before takeoff and raise them in the slipstream during flight. It was like being in an aerial motorboat.

The St. Petersburg–Tampa Air Boat Line had convenience going for it, since flying across the bay took twenty-three minutes versus two hours to drive between the two towns. But with the end of the Florida tourist season, passenger demand dwindled and the company went out of business.

Even as Tony Jannus shuttled passengers around St. Petersburg, Florida, a pilot one month younger was testing a vastly larger airliner at St. Petersburg, Russia. This prodigy was early aviator and brilliant designer Igor Ivanovich Sikorsky. His imagining of flight's future had the train for its paradigm, not the motorboat.

Sikorsky found willing believers in this vision at his place of employment because the Russo-Baltic Carriage Factory specialized in railcar manufacture. Directed to a new task by Sikorsky in 1913, the

Although designed as an airliner, the Sikorsky Ilya Muromets of 1913 was produced as a bomber during World War I.

factory's skilled workforce proudly built and rolled out the astonishing Ilya Muromets, whose record flight between St. Petersburg and Kiev is described in an earlier chapter.

The Ilya Muromets had lavish appointments. Heat exchangers on its exhaust pipes warmed the cabin, and a wind-driven generator provided electric lighting. Reflecting its proposed role as an airliner, the plane had a fully enclosed cockpit with dual controls. A door separated it from a small passenger salon with wicker chairs and picture windows.

At this compartment's rear was a small staircase to the upper bridge, as the top of the airplane was euphemistically termed. Yes, so slowly did the Ilya Muromets fly that people could walk around on top of the airplane in flight. Behind the stairs, the tapering fuselage offered a cozy private cabin with berth, cabinet, and writing table. There was even a washroom. Anyone used to first-class rail accommodations would have felt right at home aboard this airplane, although few might have ventured out on its exposed upper deck.

The hopes Sikorsky and his colleagues held for their airliner were shattered by world events. World War I preempted Russia's commanding lead in this pursuit, turned the Ilya Muromets series into a line of World War I bombers, and ultimately destroyed the imperial Russian Empire itself. One wonders what might have happened if history had taken another course.

Igor Sikorsky fled the October Revolution in 1917. Following two years in Paris, he arrived penniless in the United States, where he soon gave Pan Am its first ocean-conquering flying boats and then placed the helicopter into volume manufacture for the first time.

Sustained passenger airline operations emerged in Europe immediately after World War I. However, the air travel experience in 1919 ranged from uncomfortable at best to terrifying at worst. The

Equipped with wicker chairs, picture windows, a writing table, and electric lights, the first Ilya Muromets drew obvious inspiration from luxury rail travel. The door to the airplane's dual-controlled cockpit stands open.

era's transport planes were slow, drafty, painfully noisy, palsied by fatiguing vibrations, fraught with sickening fumes, and all too subject to the vicissitudes of weather.

German, Dutch, French, British, and Italian airliners were all in service as the 1920s began. Hands down the most advanced was the Junkers F 13, history's first all-metal airliner as well as the first to employ a fully cantilevered wing. The F 13 was also then the most comfortable airplane in existence. Fully enclosed and soundproofed, its passenger cabin featured club seating for four, plush leather upholstery, and the airline industry's first seatbelts.

Junkers F 13s flew with German, Swiss, Swedish, Finnish, Romanian, Polish, and other European airlines. These rugged workhorses also flew bush services in Asia, the Americas, and other parts of the world challenged by rugged terrain and a lack of roads. In 1921, an F 13 on floats began scheduled operations in Colombia that evolved into Avianca, the oldest continuously operated airline in the Western Hemisphere.

As the 1920s began, however, most commercial transports were

Entering service right after World War I, the Junkers F 13 seated four passengers in a plush, fully enclosed cabin.

still converted wartime types that had not been designed with comfort or boarding ease in mind. This was true of France's Breguet 14 and Farman Goliath. It was also true of Great Britain's first Vickers, de Havilland, and Handley-Page airliners.

Fokker's initial commercial offering was the Fokker F.II of 1920. Like the Junkers, it was designed as an airliner from the outset. Perhaps for this reason, these two series came to dominate European commercial aviation between the wars even though many other companies—including Armstrong Whitworth, Blackburn, Bloch, Bolton Paul, Dewoitine, Dornier, Focke-Wulf, Heinkel, Latécoère, Potez, Savoia-Marchetti, and Wibault—built propeller airliners, mostly in small numbers, for the remarkably varied European air market before World War II. Although individually those past transports may be forgotten today, they comprise a strong European and British heritage in commercial aviation.

U.S. commercial aviation also started after World War I, but it emerged with airmail because the sheer scale and challenging geography of North America outstripped the capabilities of early airplanes. Even so, America too used wartime machines as airliners as the 1920s began.

After the 1918 armistice, some enterprising Americans obtained war-surplus Felixstowe F.5s and gave these British-improved Curtiss flying boats a bulged upper-deck contour that allowed them to accommodate fourteen passengers. Designated the Aeromarine F-5L and operated by Aeromarine West Indies Airways, these former naval patrol planes inaugurated commercial services between Key West, Florida, and Havana, Cuba. A second route was soon added between Miami and Nassau via Bimini.

Prohibition was then in full swing in the United States, and so in addition to carrying international airmail, this airline did land-office business carrying partying passengers headed for Cuba and the Bahamas, where restrictions on alcoholic beverages did not apply. This airline went by the wayside later in the Roaring Twenties when the U.S. government withdrew its airmail subsidies.

Buoyed by heady investment following Lindbergh's 1927 flight,

Pan American Airways began operations late that decade on a growing network that encircled the Caribbean and Central America. By 1930, these land-plane services had been extended down the west coast of South America.

Pan Am's aggressive chairman, Juan Trippe, acquired the New York, Rio, and Buenos Aires Line in 1930. NYRBA had just pioneered services down the east coast of South America with large metal-hulled Consolidated Commodore flying boats. This acquisition and other business ventures consolidated Trippe's hold on Latin America and put Pan Am into the flying-boat business.

As the American government's "chosen instrument" for international aviation, Pan Am enjoyed lavish government subsidies as well as protection from competition. In this favorable climate, the most influential and powerful airline the world has ever known spread its wings across first the Pacific and then the Atlantic.

Pacific passenger services were inaugurated in 1935 on an island-hopping route from San Francisco to Hawaii, Midway, Wake, Guam, and finally Manila. Two years later, Pan Am extended this pioneering route all the way to Hong Kong and the Chinese mainland.

While the Atlantic presented fewer challenges in terms of range, Pan Am operations there started later because of political delays in negotiating operating rights. Only in June 1939 did the airline begin serving Southampton, England, via Newfoundland and Ireland, as well as Marseilles via Bermuda, the Azores, and Lisbon.

By then, Pan Am was operating the Boeing 314 Clipper, the largest and most capable commercial airplane in service at the time. Spanning 152 feet (46 meters) and 106 feet (32 meters) long, this flying whale weighed 42 tons (38 tonnes) and was powered by four Wright Double Cyclone engines, each producing 1,500 hp.

Traveling by Boeing 314 was as exotic and redolent of romance as scheduled air travel has ever been. Adding to the overall experience, the airline adopted luxury ocean travel as its paradigm. Subtle nautical motifs and décor graced its Art Deco facilities. Flights started with ceremony worthy of a real ocean liner as the Pan Am flight crew, resplendent in full uniform, marched across the gangplank in advance of the passengers.

Embarking on the most romantic air travel of all time, passengers cross the gangplank of a Boeing 314 Clipper, the largest and most capable of Pan Am's flying boat airliners of the 1930s.

Inside, comfortably appointed compartments offered up to seventy-four seats, which converted at night to thirty-eight curtained berths. Weight considerations generally kept passenger counts below twenty-five, with occasionally as many as thirty, so there was ample space for sauntering, chatting, or settling down to read books selected from the Clipper's library.

At mealtimes, the central salon converted to a dining room where attentive stewards served gourmet meals developed by four-star hotels and prepared onboard by chefs. Fine linens, china, cutlery, and crystal completed the elegant atmosphere. Male and female passengers even had their own dressing rooms, and at rear was a special stateroom called the bridal suite.

All of this was strictly for the wealthiest of travelers as well as senior executives whose corporations placed a premium on speed. As such, it bore little relation to air travel as people understand it today.

An attentive steward serves a gourmet meal in the Clipper's main salon, which doubled as its dining room at mealtimes.

Commercial aviation within the continental United States emerged later and was considerably less glamorous than in Europe. In 1926, landmark legislation by Congress directed the U.S. Post Office to stop flying the mails with its own pilots and airplanes, and instead contract with private companies to transport the nation's airmail. The first flight in this phased transition to private operators took place on April 6, 1926, when a small Swallow biplane operated by Varney Airlines transported airmail some 425 miles (685 kilometers) from Elko, Nevada, to Pasco, Washington, via Boise, Idaho. This historic flight by Varney pilot Leon Cuddeback linked different rail systems to shave several days off the coast-to-coast delivery time of letters bearing airmail stamps.

How U.S. commercial aviation should evolve was a question of considerable import. Within living memory, American towns had lived or died depending on whether they were served or bypassed by the nation's railroads. Now Lindbergh's flight highlighted a new form of transportation, raising similar infrastructure and public policy concerns.

Fortunately, the nation's emerging air transportation system found an able architect in Walter Folger Brown, the postmaster general during the Hoover administration (1929–1933). An aviation visionary, Brown exercised the powers of his office to usher into being a rational network of airports and airways, shape the evolution of U.S. air carriers, spur the adoption of safety-enhancing technologies, and create financial incentives for airlines to carry passengers in addition to mail.

Although pilloried for not awarding airmail contracts to the lowest bidder, Brown was correct in his assessment that fly-by-night operators with patched-up Jennys and Standards could not create the U.S. air transportation system. Well-financed corporate interests alone had the resources to lay the foundations of this emerging infrastructure.

Autocratic by nature, Brown had his way. Despite claims of favoritism and collusion, no evidence has emerged that he accepted a dime to sway his decisions. If his high-handed methods rankled others, his intentions were honorable and his actions have been of lasting benefit to the nation.

Before Lindbergh's flight, one would have been hard-pressed to buy an airline ticket in the United States. Even when flights could be found, delivery of the nation's airmail took priority, and passengers risked being evicted short of their destinations if too many mailbags waited at the next stop. As a result, at a time when European air travelers enjoyed enclosed cabins, the few Americans who flew in the mid-1920s traveled in drafty discomfort hemmed in by bags of mail.

Mail planes did not necessarily make good airliners. The Post Office had used war surplus de Havilland DH-4s. Slow, heavy, and expensive to operate with their outdated 400-hp Liberty engines, these mail planes also had just a single seat. Newer types arriving on the scene in the 1920s were not necessarily better. Some, such as Varney's Swallow or the Ryan M-1 monoplane of 1926, were smaller and thus less expensive to operate. Provided there wasn't too much mail, they could also carry a single passenger in the front cockpit, although that was clearly no basis for a successful airline.

As airplanes became larger and more powerful, dedicated mail planes remained single-seaters. The Douglas M-2 mated a big new

Commercial aviation in the United States began with mail, not passengers. The U.S. Post Office operated its own fleet of de Havilland DH-4s before commercial operators took over starting in 1926.

airframe to the surplus Liberty engine. The Curtiss Carrier Pigeon II was even beefier, a cavernous brute of a biplane with a 600-hp Curtiss Conqueror. The operators of these costly single-seaters were entirely dependent on government subsidies and could not have carried passengers even if they had wanted to.

Then on July 1, 1927, the Boeing Model 40A biplane entered service with Boeing Air Transport, an airline created by the Seattle company to operate its new product. The Model 40A had a 450-hp Pratt & Whitney Wasp engine that delivered more power than a Liberty, was more reliable, used less fuel, and weighed some 200 pounds (90 kilograms) less.

This transitional airplane featured an enclosed passenger cabin with plush leather seating for two people who traveled in addition to the mail, not instead of it when space permitted. Soon thereafter,

Placed into service on July 1, 1927, the Boeing Model 40A mail plane could also carry a passenger or two for added revenue.

Boeing brought out the improved 40B-4, a slightly stretched version with seating for four and a 525-hp Pratt & Whitney Hornet.

Boeing Air Transport had won the San Francisco–to–Chicago airmail route with a bid so low that the Post Office Department and other commercial operators expected the venture to fail, so Boeing was forced to post a bond guaranteeing its operation. Instead, the airplane's new engine technology and additional passenger revenues made the service profitable from the start.

Boeing Air Transport's first passenger was Jane Eads, a young reporter with the Chicago *Herald and Examiner*. Smartly attired in travel togs with a cloche hat, winter coat, high heels, and lipstick, she climbed aboard the Boeing Model 40A biplane in Chicago for its inaugural flight on July 1, 1927. Thus began a relentless day-and-night trip reminiscent of an aerial Pony Express. Changing planes and pilots at brief stopovers, Eads' westward journey covered 1,950 miles (3,140 kilometers) in twenty-four hours and twenty minutes aloft at an indicated airspeed of 105 mph (170 km/h).

Eads kept a diary of the flight to share with her newspaper's readers.

"We are very high up now," it read. "I feel giddy and my ears are ringing with the sound of the motor. . . . The air has been choppy and the sky full of lightning."[1]

There was pleasure in observing America from this new vantage point during the daylight hours. Eads described grazing cattle, wilderness with few roads, and the spreading majesty of rugged mountains. Particularly poignant was an impossibly lonely farmhouse on the North Platte.

Eads did not hide from her readers that she found this new mode of travel something of an ordeal. "For hours I have been flying through a forgotten country," she wrote. "One becomes unbearably weary with the altitude and the rough riding."[2]

It got worse as the Boeing climbed ever higher to clear the thrusting Rocky Mountains. "This altitude is sapping my energy," she penned. "It takes real effort for me to move at all. My feet feel like 100-pound weights. I can't lift them . . . We're hundreds of feet above the highest mountain in sight, and we're hitting some real air pockets, the kind that make your insides turn a handspring. Maybe you think I'm not getting a kick out of it."[3]

Still, she felt genuine pride in the pioneering adventure. "According to the pilot," she noted, "I'm the first woman to cross the mountains in a plane and the first woman to fly so far in the night."[4]

Such was the pace of change that a year or so later the first dedicated passenger airliners arrived on the scene. Introduced in the late 1920s, the Fokker F VII, Ford Tri-Motor, and Boeing Model 80 were noisy and drafty, but at least they gave priority to people over mail and packages. Starting in 1930, Stinson would also introduce trimotor airliner models.

Boarding passengers had only a seat by a window, a hat clip, and an air vent. Overhead was an open rack for personal belongings. On the plus side, airplanes flew low and slow, every seat was a window seat, and passengers were issued trip maps identifying sights of interest along the route. On the minus side, the very low wing loadings of 1920s-era design meant that these airplanes were thrown around if there was any turbulence at all.

Covering the Boeing 40A's inaugural flight, Chicago newspaper reporter Jane Eads became the first commercial passenger to cross most of North America by air. Here she thanks her pilot at San Francisco following a memorable adventure that more closely resembled the Pony Express of Wild West fame than modern air travel.

In those early days before pressurized cabins, airliners slogged their way through the weather instead of flying over it. Cloudbursts and thunderstorms surprised flights en route, pelting the windows with torrents of water. Lightning flashed in nearby clouds, eliciting gasps and the occasional scream from frightened passengers. Pilots added to the alarm by altering course frequently in search of better conditions, sometimes even resorting to unscheduled landings on or off an airfield (early airliners could set down in almost any good-size field).

Even fair weather was unpleasant if fluffy white cumulus clouds filled the summer sky. Their presence indicated vertical currents that soured stomachs and turned faces progressively greener as flights ran the gauntlet of constant updrafts and downdrafts. Busy at the controls, the pilots seemed immune to what the passengers experienced.

With excellent rail service providing basic domestic travel, flying was seen as dangerous as well as uncomfortable. Only the hardiest of those with above-average means gave it a try, and it was often for the novelty as much as the need to get somewhere quickly.

Nevertheless, it heralded a new era.

Postmaster General Brown decreed that there should be three U.S. transcontinental air routes. He foresaw feeder services evolving over time around the many stops along the three *trunk routes*, forming a network that would serve the nation well. To ensure beneficial competition as airlines vied to provide service, he was careful to award these coast-to-coast operating rights to different airlines.

Brown assigned the northern route to United Aircraft and Transport Corporation (UATC), the new name of Boeing Air Transport following its 1930 merger with National Air Transport, an operator of Ford Tri-Motors between New York and Chicago. This consolidation gave the carrier a continuous network spanning the continent. Pacific Air Transport with its West Coast network was now part of UATC, as was Varney Airlines. Today's United Airlines traces its history to that first commercial airmail flight by Varney's Leon Cuddeback in April 1926.

Brown awarded the southern transcontinental route to American Airways, the new name for Aviation Corporation as of 1930. That holding company had included Colonial, Universal, Southern, and other early carriers. American Airways would change names one last time to become American Airlines in 1934.

The central route—the third and final U.S. coast-to-coast route—went to Transcontinental & Western Air (TWA), a carrier created through a forced merger engineered by the postmaster general. This shotgun marriage combined Transcontinental Air Transport (TAT), a well-financed outfit that flew most of the way west from the East Coast, with Western Air Express, an energetic carrier profitably transporting mail and passengers between Los Angeles and Salt Lake City.

TAT was the first carrier to deliver passengers from one coast to the other, but it did so only with the help of two railroads. This 1929 plane-train service began in New York City with travelers boarding a Pennsylvania Railroad train for an overnight trip to Columbus, Ohio. The following morning they boarded a Ford Tri-Motor that winged its way westward to Waynoka, Oklahoma, via Indianapolis, St. Louis, Kansas City, and Wichita. From Waynoka, a Santa Fe Railroad train took them overnight to Clovis, New Mexico, where they boarded another TWA trimotor flying to San Francisco via Albuquerque; Winslow, Arizona; and Los Angeles.

With stops in all these places, coast-to-coast travel took forty-eight hours. Arduous as it was, nevertheless it saved a bit more than a day versus taking the train. The fact that this airline could not link the continent entirely by air highlighted the limitations of aviation, leading people to joke that TAT stood for "take a train."

This would change with the new decade. TWA, United, and American all inaugurated all-air transcontinental services before the milestone year 1930 was out.

Ellen Church was an Iowa farm girl working as a nurse in San Francisco. In love with aviation, she took flying lessons and also worked on aero engines in her spare time. At the airport one day, the young woman entered the offices of Boeing Air Transport and asked to be hired as a pilot. Undeterred when that suggestion was rebuffed, she next proposed that the airline hire registered nurses to attend to its passengers in flight. Both the professional training of the nurses

Aviation's first stewardesses flew with United in 1930. Other airlines quickly followed suit and public perceptions of air travel changed for the better.

and the fact that they were women would implicitly reassure nervous travelers, she pointed out.

Boeing Air Transport decided to give this idea a try, and twenty-five-year-old Ellen Church found herself leading a group of eight young, unmarried nurses who became history's first stewardesses. On May 15, 1930, aboard a Boeing 80A trimotor biplane, Church crewed the first airliner ever to carry a stewardess.

At that time, copilots generally left their seats to pass out box lunches or reassure nervous passengers. Occasionally there was a male steward. The addition of dedicated and professional female cabin staffs worked wonders psychologically. Passenger comfort improved and the experience of air travel became at once safer and more inviting in people's minds. Other airlines followed suit with female flight attendants of their own, although the requirement that they be nurses quickly fell by the wayside in light of this public acceptance.

Before there were stewardesses, there was occasionally a male steward, as on this Fokker F VII trimotor of the late 1920s.

The Boeing 247 of 1933 was sleek and modern, but Boeing did not get its interior right. Passengers disliked having to step over the front and rear wing spars that invaded the cabin, blocking the ten-passenger airliner's aisle in two places.

Douglas had a better idea. It designed its competing series with the fuselage set higher relative to the wing, giving the DC-2 and DC-3 uninterrupted aisles. More than just creating a more inviting environment, it made it easy for the stewardess to deliver meals, dispense chewing gum to clear ears as the airplane climbed, and otherwise see to the comfort of her charges.

The versatile DC-3 flew day and night and served any route with equal ease. With just two stops, it could cross the United States in fifteen hours eastbound or seventeen hours westbound. Greater wing loading gave it a better ride than its predecessors, although it too was highly subject to turbulence. Operations in bad weather, particularly winter icing conditions, made for white knuckles in both the cockpit

The airline industry came of age with the Douglas DC-3 of 1936. Each DC-3 had one stewardess and up to twenty-one passengers.

and the cabin. Nonetheless, the DC-3 put aviation on the map. With its entry into transcontinental service, the public came to view flying as the premier way to traverse the nation.

The next great leap in passenger comfort was the Boeing 307 Stratoliner, history's first pressurized airliner. A technological wonder, the 307 cruised above most of the weather at 20,000 feet (6,100 meters). Anvil-shaped thunderheads often went higher than that (indeed, some tower far higher than any jetliner flies today), but Stratoliner crews could see and steer clear of those. Here was an airplane built to cruise in sunny or starry splendor even when the earth below was enveloped in clouds.

From the passenger's perspective, the Boeing Stratoliner fundamentally redefined air travel. Derived from the Boeing B-17 Flying Fortress, an Army Air Corps bomber with turbo-superchargers for high-altitude operation, the 307 mated that military type's wings, tail, and landing gear as well as many of its systems to an entirely new and bulbous pressurized cabin nearly 12 feet (3.7 meters) across.

The result was a spacious, snug, and soundproofed propeller airliner that was smooth and comfortable in flight. And with two stewardesses catering to up to forty-four passengers, it was also the first land airliner to carry more than one flight attendant. This amazing airplane should have been a huge commercial hit for Boeing, but World War II came along and production ended after just ten were built.

For the second time in a quarter century, the demands of a global war accelerated the development of flight technologies. When this conflict ended, a new generation of transport planes far more capable than anything seen before the war had taken wing.

Douglas, Lockheed, and Boeing all brought out civil airliners based on four-engine, tricycle-gear transports they had developed during World War II. These transports would equip airlines around the world except behind the Iron Curtain, where the Soviet Union maintained an active and capable aircraft industry.

The United States then predominated in commercial aviation. Great Britain had concentrated on building fighters and bombers during the war and relied on America for most of its transport planes, as did its dominion countries. In the decades after the war, it brought out many airliner types but found success elusive in competition with the U.S. aviation industry.

As for France and other European nations, they had seen their Nazi-appropriated industrial bases shattered along with Germany's, dictating a protracted halt to airplane design and development on the continent. Fortunately, this activity has rebounded with the rise to global prominence of Airbus, a collaborative enterprise drawing on the resources and talents of many nations.

Douglas brought out the DC-4, an airliner version of its wartime C-54 transport. It followed that unpressurized type with the stretched and pressurized DC-6, the even longer DC-7 with more powerful engines, and finally the DC-7C Seven Seas, with wing-root extensions that gave it more room for fuel and pushed the engines farther out on the wing to reduce noise and vibration levels in the cabin.

Lockheed, meanwhile, had emerged from the war with perhaps the most beautiful propeller airliner of all time. The Constellation, with its triple tail and greyhound lines, was pressurized from the outset and offered speed and range that Douglas was hard-pressed to match. The DC-7 and late-model "Super Constellations" both used the Wright R-3350 Double Cyclone, whose unreliability hampered their success.

Last to market with a long-range commercial transport was Boeing, which in the late 1940s introduced the largest, heaviest, and most powerful piston airliner of all. The Boeing 377 Stratocruiser was the only commercial airliner to use the Pratt & Whitney R-4360, the largest piston engine ever produced, which suffered chronic reliability and maintainability issues.

The Stratocruiser employed technology developed for the pressurized Boeing B-29 Superfortress, which had R-3350 engines and was the most advanced airplane to emerge from World War II. At the request of the Army Air Forces, Boeing had given the B-29 a fatter fuselage during the war to produce a pressurized transport plane prototype called the XC-97.

Right after the war, Boeing came out with a considerably improved B-29 with R-4360 engines. So extensively revised was this version that the military gave it the new designation B-50. In parallel with that improved product offering, Boeing incorporated the same changes into its C-97, creating a cargo and air-refueling plane the U.S. Air Force would use for decades. This became the basis of Boeing's postwar airliner.

The capacious fuselage of the Stratocruiser created an unexpected headache for Boeing. Unlike its earlier airliners with their straightforward interiors, this one was so large that the company had more questions than answers. What should the seats be like and how should they be arranged? What amenities and appointments would postwar air travelers expect? Where would the cabin crew work and how should their equipment function? What reactions did different colors, materials, and shapes evoke? Were there dos and don'ts to interior design one should be aware of? It was a matter of ergonomics and human psychology as much as engineering.

For help, Boeing turned to Walter Dorwin Teague, one of America's premier industrial designers.

The U.S. industrial design movement arose in the 1930s to bring order to a world assailed by fast-changing technology. Influenced by streamlining and Art Deco styling, this school's leaders—Norman Bel Geddes, Raymond Loewy, Henry Dreyfus, and of course Teague—applied the creative imagination of the artist to a spectrum of products ranging from box cameras to cash registers to telephones. In the process of making them better, they defined the clean, progressive look of the decade.

Boeing's request presented Teague and his staff with an unprecedented challenge. The mission of industrial design was to bring simplicity and beauty to manufactured goods, and in the process render them intuitive to use—a quality Teague called *evident rightness*. Could this be done with an airplane?

The Teague team moved to Seattle and labored side by side with Boeing engineers for six months. The result was an interior that set new standards of elegance for air travel. In that era before high-density seating, the luxurious cabin offered large seats that converted at night into lower berths. Above were Pullman-style upper berths that folded down to open. All of these were curtained off, providing passengers with privacy after they had brushed their teeth and changed into pajamas in the plane's large gold-hued dressing rooms.

For those wishing to stay up and talk, a spiral staircase wound down to a lower lounge complete with bar, windows, a mirrored end wall, and wraparound seating for up to fourteen people. So popular was this cocktail lounge that it kept Stratocruisers in service long after their high operating costs might otherwise have warranted.

Beginning in the late 1940s, airlines introduced coach services as an economical alternative to first-class travel. High-density seating let airlines carry more travelers. While it brought lower ticket prices and a relative democratization of flying, it also made luxury

At night, Boeing 377 Stratocruiser passengers retired to Pullman-type berths and were soon lulled to sleep by the reassuring drone of powerful engines.

travel the exception rather than the norm. Configured with luxury seating and berths, the Boeing Stratocruiser carried about fifty-five passengers; in day-plane configuration with high-density seating, this same airplane carried a hundred or more passengers.

Starting with the Convair 240 and Martin 202, smaller piston airliners had also come to market for use on short routes. These postwar twin-engine airliners, and the four-engine Vickers Viscount and Lockheed Electra turboprops that followed, never sold in great number because so many military derivatives of the DC-3 (now converted back to civil use) soldiered on in airline feeder services. While those

Gold-toned dressing rooms were a hit with Stratocruiser passengers.

newer types offered higher performance and greater comfort, the low acquisition and operating costs of war surplus C-47s kept them popular with operators for decades.

The commercial jet age began with airplanes such as the Comet, 707, or DC-8. All of them were noisy to people on the ground but not to passengers inside. Compared to piston airliners, in fact, those first-generation jet transports were remarkably quiet inside and largely free of the fatiguing vibrations that had previously characterized air travel.

The manufacturers took this transition as an opportunity to redefine the passenger experience. Jet-age seats were mounted on tracks so airlines could reposition them as needed to meet changing use requirements. This in turn dictated repositionable *passenger service units,* those panels that gather together in one place each seat's reading light, fresh-air vent, and attendant call button.

When the 707 and DC-8 entered service, many airlines gave them all-first-class interiors. Premium-fare passengers flew aboard these new jets, while those traveling on economy tickets went by propeller airliner (many years would pass before the propeller fleet was entirely phased out of international service). This further consolidated in the public mind the superiority of the jet travel experience.

In 1970, Boeing took the world's breath away with the Boeing 747 jumbo jet. With a roomy cabin nearly 20 feet (6.1 meters) across, it introduced a new paradigm for air travel: the widebody jetliner with two aisles, not one.

Like the 707 and Stratocruiser before it, the 747 was the result of a creative cabin design collaboration between Teague and Boeing. Among their many innovations, this design team moved the jet's galleys and lavatories away from the sidewalls and into the center to create islands every so often down the cavernous fuselage. This distributed the passengers' seating among areas with more room-like proportions, avoiding the unpleasant tube effect that made late versions of the much-stretched DC-8 so unpopular with passengers.

Another clever design innovation was to set the 747's windows, which were no larger than those of the 707, in wash-lit reveals that registered subliminally as far larger windows, suggesting an open and light-filled environment. Experience with successive generations of airplanes has further increased the global aerospace industry's collective understanding of human psychology and the factors that influence one's perceptions of personal comfort.

Almost as wide as the 747 is the 777 twinjet, which debuted in service in 1995. The 777 introduced a curvilinear interior design that was another product of the longtime Boeing-Teague industrial collaboration. Open and spacious, it uses sweeping curves and clever indirect lighting to evoke the graceful flight of birds.

Introduced early in the 1970s, the Anglo-French Concorde supersonic transport (SST) provided a very different, highly memorable flight experience. Although the Concorde's leather seats were richly upholstered, they were surprisingly tight, and the cabin itself was cramped, with just two seats on each side of a constricted aisle. The windows were ludicrously tiny compared to those of any other airliner, and the exit door was so small that taller passengers tended to hit their heads when deplaning.

Nevertheless, the Concorde shaped positive impressions because everything from one end of this rakish machine to the other bespoke all-out performance. If one needed reminders in flight, a screen at the front of the one-hundred-seat cabin displayed the SST's current airspeed and altitude as well as the current outside air temperature.

Concordes cruised at altitudes approaching 60,000 feet (18,288 meters), half again higher than subsonic jetliners. The curvature of the earth is visible at that extreme altitude, where the outside temperature generally hovers around -75°F (-60°C). Even so, a supersonic cruise speed topping Mach 2 would heat the SST's skin through friction to the point where the Concorde's windows felt noticeably warm.

Most startling of all was arriving at the other side of the North Atlantic just three and a half hours after departing. Seat size ultimately didn't matter because there wasn't time to become uncomfortable on the Concorde. Today it's all a fading memory, however, because supersonic flight is too energy-intensive. A Concorde with one hundred passengers consumed about as much fuel crossing the Atlantic as a 747 with four hundred passengers.

Increased competition, rising costs, and reduced profit margins have taken an inexorable toll on the passenger experience in recent decades. In a sense, airlines are a victim of their own past success because people often have unrealistically low expectations of how much air travel should cost. As a result, in-flight meals and the other services and amenities that air travelers once took for granted are often no longer there.

Airlines must balance their costs against the services their passengers desire. Safe transportation to one's destination is the basic service, of course, and on short flights where comfort is less important it might be the only service. But that's not the only way that air carriers define passenger perceptions.

As it turns out, airlines also determine what kind of seats their passengers occupy in flight and how many seats in all are in the airplane. Seats are buyer-furnished equipment, and airlines obtain this BFE from third-party vendors and have them installed in their airplane by the manufacturer that is building it for them.

Seat width and especially seat pitch (how closely spaced the rows are) have a significant impact on how we as travelers feel about the trip. It's an important decision requiring a fine balance. More seats translates directly into greater potential revenue for airlines but also means less legroom and lower overall comfort for passengers.

Stuff too many seats in your airplane and your unhappy passengers will badmouth your airline to friends, hurting your reputation, even as they take their business to your competitors. On the other hand, provide too much comfort and you won't earn as much, placing you at an economic disadvantage.

In addition to the seats, airlines also decide on the in-flight entertainment system, another BFE item. With so many experience-defining decisions in the hands of the carriers, the same airplane type can exhibit very different configurations and comfort levels depending on which airline's fleet it is in, whether it's being used for short flights or long ones, and so on. While the airplane's designers determine a lot, therefore, they can't take all the credit—or blame.

If economy-class travel is less enjoyable today, the same is not necessarily true of premium-fare travel, particularly at long range. Flying internationally in a first-class cabin today can be very luxurious. Passengers typically enjoy excellent service, ample legroom, solicitous flight attendants, and gourmet foods and wines. At night, particularly on very long flights, they might occupy the latest in cocoon-style sleeper seats that convert into semi-private beds complete with adjustable reading lamps and personal entertainment systems.

With two full passenger decks, the Airbus A380 superjumbo—the world's largest commercial transport—carries more passengers at one time than any previous jetliner. The A380 is the only jetliner with two full-length passenger decks and stairs at both ends. As air travelers gain experience with this flying giant, it too will shade collective perceptions of what it means to fly in the modern world.

Many international airlines with long nonstop routes offer first-class seating that converts to flat beds, like these in an Airbus A380 superjumbo.

Airbus provides airlines tremendous flexibility to configure their A380s in ways that distinguish their service offerings and burnish perceptions of their brands. Whether or not this program is ultimately a success for Airbus and its parent company, the European Aeronautic Defense and Space Company (EADS), the airplane itself is truly remarkable.

Today the Boeing 787 Dreamliner is poised to again change our expectations of air travel. This airplane is as different inside from previous airliners as the 747 was when it entered service. Boeing and Teague expect the 787 to instill a newfound connectedness with the elements that rekindles the wonder of flight.

13 SYSTEMS INTEGRATION

MAKING FLYING SAFER

Engineering is a great profession. There is the satisfaction of watching a figment of the imagination emerge through the aid of science.

—HERBERT HOOVER (1874–1964)

Just for fun, imagine that Wilbur Wright, Louis Blériot, the Baroness Raymonde de Laroche, Igor Sikorsky, and other flight pioneers are magically transported from the year 1910 to the present day.[1] Imagine too that this party from aviation's past materializes beside a jetliner on an airport flight line and you're the only one around to explain it to them.

As their shock abates, they begin asking questions. While they have no trouble recognizing the machine before them as an airplane, they are astounded by its size and sophistication. Their wonder increases as you convey a rough idea of its speed, range, cruise altitude, and other capabilities.

Leading them around the jet, you point out its wings, stabilizers, control surfaces, and retractable landing gear. Its features and configuration make sense to them. The lack of propellers throws them at first, but they accept what you tell them about jet engines and how they work. So far so good.

Climbing the boarding ramp, you lead them into the passenger cabin, where a sea of empty seats speaks to the machine's role. Turning forward, you usher them into the flight deck. The instrument panel's glowing displays mesmerize them. Your explanation of those screens doesn't entirely register, but it's clear they understand the idea of instruments, know what the flight controls do, and grasp the concept of two-way radio communication. The gulf between their time and ours doesn't seem so great after all.

But when you mention electronic computers and the vital roles they play, puzzled frowns break out. Those frowns only deepen as you press ahead trying to explain. Blériot's eyebrows lift in Gallic disbelief. But it's when you speak the word *software* that things fall apart.

This delegation from the past is now utterly confused. Groping for an analogy they will understand, you explain that modern jets can be topped off with fuel, oil, and hydraulic fluid. However, without a magic fourth fluid—this invisible substance called *software*—they can't move an inch. Starting the engines, communicating, taxiing, and flying itself all depend on software.

Even as you say this, you know it's a waste of breath. What has defeated you is one of the most fascinating things we humans do as a species. It's called *technology integration*, and it largely explains our astonishing prowess.

Since before the dawn of recorded history, people have been taking seemingly unrelated ideas—sometimes from entirely different fields of activity—and combining them to create a whole greater than the individual parts. When this happens, one plus one can add up to considerably more than two.

This is certainly true in aviation. If you look under the skin of any sophisticated civil or military aircraft, you will find a spectrum of onboard systems (flight controls, instruments, engines, fuel, electrical, hydraulics, landing gear, pressurization, and environmental controls, among others). These systems perform critical functions and contribute to the craft's overall airworthiness.

Each of these onboard systems evolved as a rudimentary mechanical capability dedicated to one function. It was also generally entirely separate from and unrelated to the airplane's other functions (discrete systems). For example, the flight control system used wires routed through pulleys and around bell cranks to deflect the control surfaces when the pilot moved the control stick and rudder pedals; this system was not connected to any other airplane system.

This began to change because human ingenuity kept coming up with new ideas and clever ways to knit them all together. The result has been profoundly transformative capabilities that make modern air travel vastly safer than engineers and pilots back in the propeller era would have dreamed possible.

The invention of the humble radio altimeter in the 1930s launched one such wave of technology integration.

Inventive genius Nikola Tesla demonstrated the world's first radio transmitter in St. Louis, Missouri, in 1893. Although Lloyd Espenscheid was only a small boy at that time, that event in his home town proved formative; he grew up fascinated by electricity.

Sent off to Brooklyn, New York, at age twelve to live with family friends, Espenscheid took up ham radio and telegraphy. He put these skills to use during summer vacations, earning extra money as a wireless operator on ships sailing out of New York. Leaving college with a degree in electrical engineering, Espenscheid helped install cutting-edge radio equipment aboard U.S. Navy vessels before signing on with American Telephone & Telegraph in 1910.

Over the next four decades, Lloyd Espenscheid performed groundbreaking research as a scientist in AT&T's engineering department. One result was a 1924 patent for a railroad collision-avoidance system. Its purpose was to provide timely warning of another train on the tracks ahead by emitting radio waves and sensing their bounced return. Another patent arose from a more sophisticated version of this invention turned vertically to serve pilots.

Espenscheid had come up with aviation's version of the sounding line, a weighted cord lowered into water to ascertain its depth. That maritime practice, which dates back to ancient times, had in the latter nineteenth century provided American journalist, humorist, and satirist Samuel Clemens—one of America's greatest writers—with the pen name Mark Twain. It was a term recalled from Clemens' days as a Mississippi riverboat pilot on the eve of the U.S. Civil War. The riverboat's leadsman would lower a plumb line to take soundings of the river's muddy depths. Knots tied in this rope marked off the depth in fathoms. A cry of "mark twain" denoted two fathoms (12 feet or 3.7 meters) of water beneath the keel, and that meant safety.

Espenscheid's electronic counterpart for aviation was very different from Paul Kollsman's barometric altimeter. Whereas the latter device showed the airplane's height above sea level, Espenscheid's radio altimeter bounced a signal off the surface below instead to tell pilots how high they were above the ground. This was potentially valuable information at night or in bad weather.

Astonishingly precise time measurements would have been required to determine height by actually timing the echo. However, an easier alternative existed that worked just as well. This was to transmit a continuous downward signal that oscillated up and down in frequency at a set rate, creating the radio-wave equivalent of a siren's ululation. The greater the distance from which this wavering signal bounced back, the more the frequency of the airplane's transmission would have shifted in the interim. Because comparing two frequencies is a simple matter, Espenscheid employed this frequency modulation approach.

The Western Electric Company—AT&T's manufacturing affiliate—marketed the first commercial radio altimeter in 1937. The more accurate term *radar altimeter* came into use following World War II (*radar* is an acronym for *radio detection and ranging*).

In modern flight decks, the radar-altimeter readout appears on the *primary flight display* screens in front of the pilot and copilot. These displays show at a glance the airplane's attitude, speed, altitude, autopilot mode, and a wealth of other flight-instrument information.

During the approach and landing phases, the radar altimeter also provides inputs for a computer-generated voice that calls out the decreasing altitude in increments as the airplane nears the ground.

Flight crews use these automated announcements to know when to *flare*, which means to pull back on the control wheel and moderate the contact of the wheels with the runway. While pilots can also gauge this visually, the precise readouts help them land more consistently.

People soon got the idea of leveraging radar altimetry to serve other purposes. The first additional use came in the late 1960s with the introduction of the *ground proximity warning system* (GPWS), which alerts crews to decreasing separation between the airplane and the ground.

Analysis of commercial airliner accidents showed many to be the result of *controlled flight into terrain*. These CFIT (pronounced "SEE-fit") accidents were often the result of a loss of situational awareness, fatigue, or fixation on minor systems failures and other distractions by the flight crew. The airplane itself was fully airworthy and under control at the time of the accident; had the same circumstances arisen in conditions where the ground was visible, a crash would not have ensued.

When triggered by the airplane descending to the surface below, the ground rising up to meet the airplane, or a combination of the two, a "ground prox" alert fills the flight deck with loud aural tones and strident words. Depending on the specific warning issued, flight crews might hear "Terrain, terrain," "Pull up, pull up," "Sink rate," "Glide slope," and so on.

Thanks to the introduction of GPWS, the number of CFIT accidents fell dramatically, but they still occasionally occurred. Particularly disheartening were those cases in which, as confirmed by the cockpit voice recording, the flight crews had willfully ignored what should have been a timely warning.

Part of the problem was that, like the proverbial boy who cried wolf, early GPWS systems were plagued with false alerts. But there was more than that going on, and experts from across the global commercial aviation community—airlines, pilots, manufacturers, government regulatory authorities, and other interested parties—convened to determine what it was.

This campaign against CFIT accidents was pursued on two fronts. One was GPWS itself as manufacturers fine-tuned the trigger-threshold algorithms (an algorithm is a series of defined mathematical steps) to reduce false alerts. While improvements were made, however, these systems could not achieve foolproof results because a radar altimeter only looks downward, not forward, and mountainous terrain can of course rise dramatically.

Even as those efforts progressed, multidisciplinary teams were probing why professional flight crews might choose to discount GPWS warnings. These human-factors studies examined the mental models that we human beings construct and maintain of the ambient situation as we understand it. They highlighted how persuasive these mental models can be and how, when they are at odds with reality, they can blind us to what would otherwise be evident. For example, a flight crew might be so convinced that the autopilot is holding altitude that they fail to realize it has been accidentally disengaged and the instruments now show a descent.

Based on these efforts, airline training was universally implemented with a revised procedure for responding to a GPWS alert. This was to climb out immediately in an act-first, ask-questions-later response. Thanks to these and other improvements, CFIT accidents in the world commercial jetliner fleet are largely a thing of the past.

But people weren't done building on Espenscheid's simple invention. This collective activity would transform air travel and further enhance safety.

Starting in the late 1970s, the United States launched into orbit a constellation of satellites known collectively as the NAVSTAR Global Positioning System. Maintained by the U.S. Air Force, this infrastructure is available to the entire world. A minimum of twenty-four operational GPS satellites (actually just over thirty at the time of this writing) girdle the earth in highly inclined orbits that ensure excellent coverage of the entire world.

GPS is freely available to aircraft, ground vehicles, ships, soldiers, scientists, hikers, and any other users worldwide. This system provides accurate positioning to within a meter (about a yard) for civil users. With atomic clocks aboard all the satellites, GPS also provides astonishingly accurate time information.

Other satellite navigation systems are also being fielded or planned, notably Russia's GLONASS and the European Union's Galileo systems. In addition to those global constellations, China and India have regional systems in the works. For simplicity's sake, therefore, the pioneering GPS and those that follow are collectively referred to as the *global navigation satellite system*.

Aviation relies heavily on this GNSS infrastructure. Jetliners flying the world's oceans can be certain of their position despite the lack of positive radar coverage at sea. And where GNSS is augmented regionally by one or more ground-based transmitters, it becomes even more accurate, allowing civil aircraft to fly precise instrument landing approaches without the need for traditional instrument landing systems located at the airport itself.

In the 1930s, airliners such as the DC-3 used a radio-navigation system known as the *four-course radio range*. Depending on whether the airplane was to the left or right of the intended course, its flight crew heard the Morse code for either A (._) or N(_.) in their headphones. When on course, these two signals merged together to become a constant tone (later on, a needle on the instrument panel displayed these indications, sparing one of the pilots from having to listen for hours at a time).

This system allowed airliners to follow invisible electronic airways in the sky or perform instrument approaches and landings at properly equipped airports. In the latter case, the crew followed published approach procedures involving timed turns and required headings and altitudes. Vertical fan markers, audible when the airplane passed directly over them, marked thresholds bracketing a constant-rate descent to the runway.

If the crew sighted the runway by the time a designated minimum altitude was reached, they went ahead and landed. If not, they declared a missed approach and started all over again or headed to an alternate airport to try their luck there. It was all highly imprecise and extremely stressful and fatiguing for pilots.

With the discovery of radar in World War II came the ground-controlled approach. This military system had trained controllers "talk crews down" (verbally guide them to the runway) through bad weather by reference to two screens. One showed the runway's extended centerline, while the other showed the desired descent angle. Although effective, ground-controlled approaches soon gave way to better systems.

The key evolution in the postwar era was the instrument landing system (ILS), in which flight crews follow vertical and horizontal guidance bars to track the localizer (extended runway centerline) and descent angle (glide slope). The precise ILS serves the global industry well but is fundamentally complex and expensive. As such, it is not available at the vast majority of the world's airports. Even at those airports large and busy enough to have ILS, it is generally not available on every runway.

Where ILS is not available, pilots have found themselves back to using nonprecision approaches involving a variety of radio-navigational aids. Right up to the present day, air transport pilots are required to remain proficient in these old-fashioned "step-down approaches" as a fallback in the event that ILS is unexpectedly unavailable.

Satellite navigation is mercifully spelling an end to those "sweaty-palm" nonprecision approaches that look back to aviation's earlier days. It is also providing an alternative to costly radar and ILS systems. Instead, with very little added costs, virtually any airport in the world can now have precision approach and landing capabilities collaboratively developed by all the participants in the global aviation system. These stakeholders include regulatory authorities and other government agencies, airlines worldwide, aerospace manufacturers (the makers of airplanes, engines, and equipment), and interested nongovernmental organizations (pilots' unions, passenger organizations, and safety and other advocacy groups).

GNSS allows safe, precise, and flexible approaches that avoid the surrounding terrain of even the most challenging airports. These systems preempt the need for conventional air traffic control based on positive radar coverage. This is great news for developing nations, which are spared the heavy investment burdens of developing conventional aviation infrastructures.

Under the coming air traffic management paradigm, jetliners and other aircraft will automatically report their identity, GNSS-derived position, heading, altitude, vertical speed, and other pertinent information at frequent regular intervals. This automated data-burst reporting, a programmed function of the airplane's flight-management computer, draws needed information from the airplane's air-data computer and employs satellite communication data-link capability.

Transitioning from air traffic control to air traffic management will shift much of the work of aircraft separation to the airplanes themselves within this integrated, information-rich environment. Ground-based controllers will intervene only as needed to resolve arising conflicts if two airplanes are headed for the same airspace. Instead of traffic bunching up along traditional airways and over electronic intersections, flights in this coming environment will be spread out for more efficient use of available airspace.

Already under way, the phased transition to this future air traffic environment will take decades. Major systems integration challenges must be met, enabling technologies must be developed and validated, political considerations such as cost and airspace allocation must be resolved, and a spectrum of other issues and concerns must be properly addressed.

For example, how can the world transition with uncompromised safety from classic air traffic control to the future air traffic management environment when during this interim period some airplanes will be capable of air traffic management and others won't? And what of the very light jets (formerly termed microjets) now entering service, which raise concerns at a challenging time because they combine business-jet performance capabilities with single-pilot, privately owned and operated general aviation usage patterns?

In recent decades the world has witnessed exponential growth in computer processing speeds and available memory. These astonishing technological capabilities have in turn seen three-dimensional computer applications of all kinds come to market. From action-adventure games to flight simulators to home-design tools, low-cost software has opened up an astonishing variety of virtual worlds to us on our personal computers.

Aviation too drew on these newfound capabilities for yet another safety-enhancing tie-in with the lowly radar altimeter. The idea here was to combine an electronic library of digital terrain data with GPWS for enhanced ground-proximity warning. Together with GNSS, this new system is little short of revolutionary.

Called TAWS, for *terrain awareness and warning system*, this mandated safety system combines the radar altimeter's downward sensing with knowledge of the surrounding terrain as well as the airplane's position, course, airspeed, vertical speed, accelerations, and configuration (gear and flaps). This gives it the ability to anticipate what's ahead for more informed and accurate ground-proximity alerts. In mountainous areas, it can also allow the system to guide the crew safely around the highest ground.

Although often described as a forward-looking alerting system, TAWS does not actually sense what is ahead of the airplane. Moreover, its digital terrain data set may fall out of date over time—for example, if a large transmission tower is built atop a hill near an airport. Consequently, provision is made for periodic updating of this digital electronic library.

This virtual world also has the potential to be shown on head-up displays, which are angled glass panels on which projected information, focused to infinity, overlays what a pilot sees ahead. Once found only in military aircraft, head-up displays are today widely used in commercial jetliners. They repeat flight instrument data from the panel and display overlaid symbology offering approach and landing guidance.

Projecting digital terrain data is possible on a head-up display, and it would fill in the missing outside world when flying in bad weather. However, this use of synthetic vision—so called because it is merely a representation of the airplane's surroundings, not an actual view of those surroundings—has the potentially dangerous drawback of looking persuasive while failing to represent objects that might actually be in the sky.

Head-up displays can also show infrared or other alternative-spectrum imagery to jetliner flight crews. Called *enhanced flight vision systems*, capabilities of this type are being developed and tested.

Unlike synthetic vision, enhanced flight vision systems do depict what is actually out there even if it is not visible to the naked eye.

Technology integration is transforming our lives, not just aviation, and the pace of change has never been greater. Where will it lead? The sky's the limit.

14 TODAY'S STATE OF THE ART

THE BOEING 787 DREAMLINER

The easier it is to communicate, the faster change happens.

—JAMES BURKE, BRITISH SCIENCE HISTORIAN AND BEST-SELLING AUTHOR[1]

Within a human lifetime, the idea of traveling between continents by air has gone from unthinkable to normal; in fact, what is unthinkable now is to travel very long distances any other way. Flights whisk us effortlessly across the Atlantic, Pacific, and North Pole, reducing those once-absolute barriers to mere hours spent aloft. Today's most capable jets can fly 10,000 miles (16,000 kilometers) nonstop; in less than a day, they can deposit hundreds of people at a time almost halfway around the globe from where they took off.

How did we accomplish this? Scientific knowledge and technological prowess are what leap to mind, but those are not the entire answer. In fact, none of it would have happened if human interaction hadn't evolved just as dramatically. As a consequence of our improving ability to communicate and collaborate, the pace of change has grown exponentially since the dawn of human flight, an accelerating trend that continues today.

The industrial revolution, early aviation's backdrop, could not have happened without the printed word. Technological advancements on this front brought the daily newspaper into being early in the nineteenth century, accelerating the proliferation of information. The more people knew, the more serendipitous and often highly productive connections that were made.

Thinking from many parts of the world solved flight's challenges, of course, but those human ideas flowed at a very leisurely pace back then. Late in the nineteenth century, a handwritten letter from Lawrence Hargrave took about six weeks to reach Octave Chanute via steamship and railroad. Chanute's reply took just as long to return from Chicago to Sydney. Time frames between Europe and Australia were about the same.

In marked contrast, scientific colleagues today would probably communicate via e-mail, a medium that allows our thoughts to flow as fast as we can type them because it has decoupled the written word from any physically transported medium. Better still, our e-mails go simultaneously to as many parties as we like, and we also have other options at our fingertips such as instant messaging and global phone service. Even more ideas and information reach us through the broadcast media and the World Wide Web.

Yes, thoughts fly vastly faster today, but that's just part of the answer, because a spectrum of new tools and processes also helps us work together more efficiently and effectively. Personal and mainframe computers let us accomplish tasks with astonishing speed and extend capabilities not previously available. Prodigious computer processing power and memory allow us to reduce vast amounts of data, perform aerodynamic modeling, and even design airplanes in three dimensions using software that also supports manufacturing innovations.

Aeronautical engineers designing airplanes work from all parts of the world, yet they collaborate as efficiently as if they were in the same room. These widely distributed teams hold virtual meetings using Web-based tools that allow each participant to view the same charts and discuss them freely. Some collaboration softwares include an inset window showing the person speaking at the time, not just the data that person is presenting. These new tools are revolutionizing how technologists collaborate. Whereas people once flocked to the work, today they let the work flow to them.

Many other trends are likewise transforming collaboration. One is ongoing technology integration, which is blurring the boundaries between formerly discrete devices. The cell phone offers capabilities including e-mail, Web browsing, instant messaging, and features such as a personal calendar, calculator, alarm clock, camera, verbal memo recorder, and so on. Some are so capable at other tasks that they are viewed primarily as personal digital assistants, with telephony as just one function.

The same is true of formerly discrete fields of activity, where technology integration is blurring the lines between formerly distinct job functions. Keeping all this straight makes for organizational challenges and requires flexible teaming that calls on skills as needed.

This ongoing integration of technologies lets engineers create highly complex systems of systems in which individual capabilities are knit seamlessly into a whole of vastly greater overall capability. The coming air traffic management environment provides an example of the potential scope of system-of-systems engineering projects.

The designs of the latest jetliners and military aircraft likewise reap the benefits of technology integration. They are aerial platforms whose performance capabilities are defined not just by airframe efficiency, aerodynamics, and engines but also by electronic sophistication. This is particularly true of fly-by-wire aircraft.

Much like your multitasking cell phone, modern airplanes rely on shared and distributed functionalities rather than the discrete systems of aviation's earlier days. This makes them more capable, reliable, redundant, and robust. They have in effect become flying data networks whose distributed avionics and other systems interconnect via digital data bases using Internet-type protocols.

As for aerospace engineering itself, its many constituent fields (aerodynamics, structures, electrical systems, payload, flight deck, landing gear, and so on) have advanced tremendously over time. More than just possessing greater knowledge, these disciplines also have vastly better tools and processes. Whatever the task—program planning, systems and component design, risk identification and mitigation, and so on—human ingenuity and collective effort have devised better ways of accomplishing it.

Engineers care enormously about definitions, standards, and baselines. They have to be sure they're talking about the same things, so terminology is carefully agreed upon and ambiguity is banished. Engineers also like to know they're using compatible tools and processes and that they're working with the same data sets and document iterations.

Technology and human inventiveness continue to transform every bit of aviation activity from one end of the field to the other. For example, engineers used to create airplanes by working toward periodic updates of an evolving design. Unfortunately, what one engineer or group did in building further on that baseline release often conflicted with what others did elsewhere, an undesirable situation that wouldn't necessarily be known until too late, when changes were either too expensive or impossible. The result was a suboptimal design people had to live with. Today, in contrast, computerized design databases update immediately, so engineers always see the latest baseline, reducing errors and rework.

Airplanes are today defined by teams that bring together not just the people who will create the design but also those who will manufacture it and those who will operate and maintain it in service. Pioneered by Boeing with its 777 program, this design-build-support team concept ensures full attention to every aspect of the design. It makes for a better-thought-out airplane that is more user-friendly.

Greater cost and performance visibility is another beneficial trend in aerospace. Because products such as commercial jetliners cost billions of dollars to develop, it is imperative to make decisions that manage program resources wisely and deliver the right product. Only the availability of sufficient information allows for informed decisions rather than guesswork. As a result, companies today are striving for unprecedented cost and performance visibility that make possible *program metrics* providing needed insight and showing actual progress.

All aerospace companies have farther to go on this front, some by quite a bit. The best major manufacturers are beginning to train their suppliers in the systems they have devised so that they too use the same metrics. In turn, many of these first-tier suppliers will do the same with the companies that they depend on, resulting in "same baseline" visibility up and down the supply chain. The ultimate goal is better decision making, further optimization, and greater success enabled by facts and data.

"I think I can build a better airplane."[2]

So declared William E. Boeing after his first airplane ride in 1915. The experience had thrilled the wealthy Seattle businessman and set him thinking. To prove he was right, he founded an airplane company in 1916.

The Boeing Company is still striving to give the world better airplanes all these years later. The latest is the Boeing 787 Dreamliner, an ultraefficient airplane that represents a bold leap forward. This

Boeing is redefining air travel with the 787 Dreamliner, whose passenger cabin is as advanced as the airplane itself.

commercial transport uses 20 percent less fuel and costs 30 percent less to maintain than similar-size jetliners. In an industry where a difference of 1 percent can mean millions of dollars annually to an airline, these improvements are nothing short of revolutionary.

But lower operating costs are just part of what's new with the 787. The rest is a literal reinvention of the experience of flying itself that will change expectations of what air travel should be. In fact, the Dreamliner's passenger cabin is as much of a leap forward as the airplane itself.

While this 787 Dreamliner bears the famous Boeing brand, it is not solely a Boeing or even a U.S. creation, as the company itself is quick to point out. Instead it is the product of a vast global collaboration whose challenges were solved in the United States and dozens of other nations.

Talented engineers on many continents participated in designing this airliner. Large parts of it are being manufactured abroad. Arriving as built-up major assemblies, these components ultimately come together in Everett, Washington, where the 787 is assembled rather than built in the traditional sense.

Ironically, much of the technology embodied in this twenty-first-century airliner was actually developed for a different airplane envisioned by the company.

In March 2001, Boeing unveiled the Sonic Cruiser, an unusually configured jetliner whose state-of-the-art technology would let it cruise at very nearly the speed of sound, shaving hours off long-distance air travel. Then came the terrorist attacks of September 11, 2001, which altered the economic landscape for airlines worldwide and hit U.S. carriers especially hard.

Responding to this changed situation, Boeing offered its current and potential airline customers a choice between the Sonic Cruiser and the 7E7, which was a parallel concept the company had pursued as a baseline to show what the transonic Boeing Sonic Cruiser's technologies could do in a more traditional application. The subsonic 7E7 would cruise at regular jet speeds but consume less fuel, produce fewer emissions, and be less costly to operate.

Hands down the world's airlines opted for the ultraefficient 7E7, which became the 787 well after program launch in April 2004. For the first time, Boeing also asked the public for help adding a name to this product's numerical designation. People the world over cast a half million online votes and the 787 became the Dreamliner.

The 787's enhanced fuel efficiency comes at a crucial time. Spurred by rising fuel costs and global environmental concerns, it has amassed the strongest early sales of any commercial airplane in history.

In a case of highly beneficial competition, Airbus is developing a slightly larger midsize jet designated the A350 XWB (the initials stand for *extra wide body*). Airbus and EADS, its corporate parent, hope that their advanced-technology offering will match if not exceed the 787's level of performance.

The Dreamliner doesn't look all that different from other jets. This low-wing monoplane has two pylon-mounted engines, one on each swept wing, which is the configuration of well over 90 percent of the jetliners that come off the assembly lines these days.

Aerodynamic optimization has pushed us to increasingly similar designs over the decades. This process in many ways mirrors natural selection, as one can see by comparing the shark and the porpoise. Although they are taxonomically entirely different, hydrodynamic optimization has caused them to evolve similar forms.

Even so, the 787 Dreamliner will not be hard to pick out on a crowded airport flight line. A distinctive nose profile and unusual cockpit window framing together make it easy to spot if one knows what to look for. Not so readily apparent are this airplane's technological innovations, which occur in four areas: airframe, engines, aerodynamics, and systems.

The 787 is the first jetliner to be made primarily of lightweight composites instead of aluminum. Virtually 100 percent of its skin is made of carbon-fiber-reinforced plastic, as is much of its internal structure. Composites are stronger and lighter than aluminum and, unlike metals, they hardly fatigue or corrode. Composites are also considerably less sensitive to damage, which is one reason why military helicopters have used composite blades for decades. Even after a lifetime of hard use, a composite 787 will still be stronger than a brand-new metal jet the day it rolls out of the factory.

Aerospace manufacturers have decades of experience with composites in structural or load-carrying applications. A good example is the empennage of the 777, Boeing's previous jetliner before the 787. The 777's tail surfaces are entirely composite. Whereas composites account for 12 percent of a 777's empty weight, however, they are 50 percent of the Dreamliner's empty weight and—composites being so light—much of its total volume.

To build 787 fuselages, Boeing engineers figured out how to manufacture large sections as single pieces. Built in Italy, Japan, and the United States, these light and durable fuselage barrels incorporate internal structure and are pre-stuffed with wires and tubing to speed assembly. Completing a fuselage involves little more than bolting these barrels together, a rapid process that requires 50,000 fewer fasteners than conventional fuselage manufacture.

The Dreamliner's gracefully curving wings are equally advanced and are in fact the most aerodynamically efficient yet fitted to a commercial jetliner. These all-composite wings represent a dramatic departure from past practice. Greater material strength let designers define a wing that is somewhat longer and narrower than in the past. This higher aspect ratio enhances overall aerodynamic efficiency.

Helping to shape the 787's wings was the latest in computational fluid dynamics, an aerospace modeling tool that harnesses supercomputers to perform millions of calculations per second according to known fluid-flow and other physical principles. Because computational fluid dynamics has capability and accuracy limits, Boeing engineers also performed extensive wind tunnel testing to validate and fine-tune the wing design.

Composite weight savings also benefited the Dreamliner through a *virtuous circle of design*. This occurs when an improvement to one design parameter yields benefits in others thanks to a cascading redefinition of what is needed to achieve the airplane design's stated performance goals. For example, reducing the airframe weight means that less fuel is needed to fly a given distance, which in turn reduces the weight of and internal volume required for fuel, for a further decrease in airplane gross weight. These readjustments in turn dictate a slightly smaller wing, which reduces aerodynamic drag and weight to further trim fuel requirements. These weight reductions in turn mean that the landing gear can be less robust and thus lighter, and so on.

Important as these benefits are, however, they are just one part of the 787's clever use of technology. And as is true with all airplane designs, it is not just the technology itself but how people combine it.

The 787 cabin must lift passenger satisfaction to a new plateau, it must reconnect travelers with the wonders of flight, and it must be as much of a leap forward as is the airplane itself. These self-imposed goals guided Boeing and Teague designers as they developed an interior for the Dreamliner, which they felt should be an antidote to stress, crowding, and long lines at airport security checkpoints.

Building on everything its experts knew, and conducting pioneering studies to learn more, the interior design team met this goal in innovative ways. The sense of a very different travel experience begins right at the door, where a spacious, bright, and high-ceilinged entryway welcomes passengers coming off the jetway. This modulated use of space marks a transition in travelers' minds, subliminally informing them that they have left their hassles behind.

Inside the Dreamliner, the spacious cabin with gently flowing curves and innovative lighting creates a sense of tranquil well-being. Heightening this perception, the interior employs the arch—an ancient symbol connoting strength and harmony—as its key theme. The sidewalls periodically constrict, suggesting archways that serve to define individual room-like areas within the airplane. Gone is the sense of a long tube, even though access remains open from front to rear. Depending on how airlines configure their 787s, galley and lavatory islands can also provide architectural boundaries.

In a case of *trompe l'oeil*, the overhead bins are much larger on the inside than they appear on the outside. Some let you slip even the largest carry-on cases in edgewise, short end first. This is possible because the fuselage's inside arc does not follow the fuselage's outer curve, resulting in an unprecedented amount of available stowage volume per passenger.

Whereas the cabin of other jets can drop to a pressure equivalent to an altitude of 8,000 feet (about 2,400 meters) in flight, the 787's cabin altitude reaches only 6,000 feet (about 1,800 meters). Making this possible is the strength of composites, which permits a greater pressure differential between inside and outside. Boeing chose this lower limit based on aeromedical data showing that the headaches and fatigue experienced by many air travelers do not occur at or below a pressure equivalent to 6,000 feet.

And whereas the air of most jetliner cabins is extremely dry in flight, the 787's cabin air is more humid to eliminate the irritated throat and contact lens distress many passengers experience. This higher humidity is possible because composites do not corrode.

Perhaps the Dreamliner cabin's most striking feature is its windows, which are larger and taller. Whereas other jetliners' windows serve just the immediate window-seat occupant or perhaps that seat row, the 787's windows are visible above the tops of the seats so that they serve the entire cabin. This design feature lets passengers gaze freely at distant horizons and enjoy enhanced connectedness with the elements. Whether viewed directly or sensed peripherally, this link with the outside world is meaningful to us as a species on a fundamental level.

Larger windows rekindle an intimacy with the sky that air voyagers once took for granted but lost for most of the jet age. And although one might think that these large windows might alarm nervous fliers, the opposite is true because they reduce feelings of confinement and give meaning to sensed motions.

If passengers like, the 787's windows can be electronically dimmed at the touch of a button. Flight attendants can also control these electrochromic windows globally to prevent bright ambient light from disrupting in-flight movies. An additional sense of connectedness with the natural world comes from the state-of-the-art lighting system, which employs light-emitting diodes in three colors to create a subliminal but persuasive sense of open sky overhead. On long

A high-ceilinged entryway welcomes boarding Dreamliner passengers with the uplifting sense that all hassles have been left behind.

flights, this washed-lighting treatment shifts in hue to nudge travelers' thoughts and biological clocks toward the new time zone.

Like city buses in relation to vans or cars, bigger jetliners are supposed to be more economical per passenger seat than smaller ones. However, the Dreamliner—a medium-size jet smaller than the 777—is so efficient that its fuel consumption per seat is suggestive of a much larger airplane. Helping to explain this astonishing fuel efficiency are the new-generation Rolls-Royce or General Electric engines that power the Dreamliner.

This airplane's high-bypass-ratio fanjets are equipped with the latest in noise-reduction technology. The jagged serrations at the back of the 787's engine nacelles make the airplane even quieter by affecting how the ambient air mixes with the engines' bypass and core-exhaust flows.

The Dreamliner also saves energy through dramatically rethought airplane systems centered on a more electric architecture. It's a very different flying machine under the skin, one that requires less

An arch motif, innovative lighting, larger and taller windows that serve the entire cabin, and greater humidity in the air also distinguish the Dreamliner travel experience.

energy to operate and does not squander it when not needed the way many current airplane systems do. This systems rethinking also makes the airplane inherently easier to keep airworthy. For example, previous jetliners constantly bleed high-pressure air off their engines to pressurize their cabins. This robs the engines of some thrust and adds to fuel consumption. Those traditional pressurization systems also require extensive pneumatic ducting complete with precoolers and check valves, all of which adds weight to the airplane and demands considerable maintenance time and attention.

In contrast, the 787's no-bleed architecture dispenses with that ducting and does not rob the engines of high-pressure air. Instead, electrically driven compressors pressurize the cabin. The result is a fundamentally simpler, lighter, and more reliable system that consumes up to 35 percent less energy and requires far less maintenance.

The Boeing 787 Dreamliner has set off a new burst of collective inventiveness in the arena of flight. While it is difficult these days to put a single human face to advancements in aviation, for those in the field, it is precisely this more inclusive, integrated, and effective process of creation—an unprecedented richness of collaboration—that makes aviation an exciting field. It should be noted that today's aerospace workers are vastly more knowledgeable and better trained than their illustrious precursors from flight's early days.

More than two centuries after George Cayley first imagined the airplane, and over one century since Wilbur and Orville Wright invented it, progress in flight continues at an astonishing pace.

Humankind's oldest dream lives on.

POSTSCRIPT

TOMORROW'S WINGS

FUTURE AIR TRAVEL TECHNOLOGIES

Our success aloft shows how much we humans can accomplish when we work together to a shared vision. This success and our accelerating technological prowess suggest that amazing things will happen when we begin collaborating broadly in other areas.

Of the many challenges confronting the world in the twenty-first century, global climate change is the most significant. Based on vast evidence and informed by our best scientific understanding, a broad consensus has emerged that global warming must be addressed effectively. Hand in hand with this global focus is the final consolidation of a long-emerging paradigm shift. This beneficial paradigm sees the planet as a closed system with finite resources rather than the boundless frontier we once thought it to be.

Aviation's future is inextricably combined with our perspective on and attention to the global environment. Ultimately, the world may need to fly less. In the meantime, aviation must "clean up its act" even as the rest of the world does.

Jetliners burn kerosene, which is the traditional fuel of turbine-powered aircraft.[1] Since the jet age began, the world has learned how to build jetliners that use 70 percent less of the stuff per passenger seat. This progress continues with the 787 Dreamliner and the new engines developed for it.

Despite this trend, growth in the number of jetliners in service around the world threatens to significantly increase aviation's total contributions to global warming in the coming years. Consequently, the industry is taking bold steps to further improve its environmental performance.

Like other transportation sectors, aviation is looking to biofuels to reduce emissions. Because the carbon dioxide in biofuels was pulled from the atmosphere when the fuel's feedstock grew, it does not represent a net addition of CO_2 to the atmosphere, in contrast to the carbon released by petroleum-based fuels, which was formerly underground.

Aviation is particularly challenging because jetliners need a fuel that packs a very high energy density and has inherent resistance to low temperatures. This eliminates ethanol as a candidate because it has a low energy density; an equivalent amount of ethanol would take a jetliner only half as far. It also eliminates biodiesel, which offers 80 percent of the energy density of kerosene but would solidify at cruise altitude, where temperatures typically hover around -60°F (-50°C).

Fortunately, energy experts are now evaluating a promising alternative: microscopic algae that have the ability to produce lipids convertible into a fuel closely resembling kerosene. These microalgae need only bright sunlight and CO_2. They thrive in brackish water unfit for other use, making harsh desert environments ideal for their cultivation. Consequently, large-scale production of bio-jetfuel would not require arable land, thus avoiding the downside of biofuels that pit the world's desire for energy against its need for food and forests.

Algae are far more productive than terrestrial plants grown for energy. Algal production in shallow ponds produces oil yields up to fifty times higher than oilseed crops grown on an equivalent area of farmland. Moreover, these algal cultures draw carbon out of the atmosphere so aggressively that their use is also being explored for scrubbing CO_2 out of the industrial effluents of refineries, power plants, and other existing point sources of industrial pollution.

Intensive cultivation in enclosed plastic tubes called reactors holds out the promise of large-volume production as well as significant economies of scale. The aviation industry is working with other interested parties to see whether long parallel rows of these reactors, or other cultivation methods, might supply aviation's future needs. Other potential biofuels are also being evaluated, as is the use of bacteria or algae that are genetically engineered to produce hydrocarbon chains. These might be the basis for a range of synthetic fuels that do not add new CO_2 to the atmosphere.

The effects of other jetliner emissions aside from CO_2 are also being addressed. Oxides of nitrogen (NO_x) are the chief focus of concern. These emissions create ozone (O_3), a potent greenhouse gas when released high in the atmosphere.[2] Broad efforts are in progress to reduce NO_x formation in jet engines. Water vapor, another emission, is only a mild greenhouse gas, but it freezes into long white contrails that promote the formation of cirrus clouds. The impact on the global environment of this cloud creation is another focus of scientific study.

Human inventiveness and technology translation will benefit aviation in many ways. One example that will soon be available is the *fuel-cell auxiliary power unit*, whose developers include Airbus and Boeing.

Fuel cells are devices that convert a fuel into electricity without combustion. They have no moving parts and are inherently reliable. When fueled with hydrogen, their only exhaust is hot air and pure water vapor. With a reformer attached, they can also run off conventional hydrocarbon fuels such as jetfuel. They do create some emissions in this case, but far fewer than combustion engines.

Before too many more years have passed, the *power density* (power output relative to weight) of fuel cells will have improved to the point where they can play efficiency-enhancing roles aboard commercial jetliners. For example, more-electric-architecture airplanes like the 787 could use fuel cells as their auxiliary power unit. This fuel cell APU in the tail cone could even provide primary electrical power in

Jet transports as we know them today may one day give way to futuristic concepts such as the blended wing-body (BWB).

flight, offloading the engine-driven generators to make these airplanes even more fuel efficient in the future.

"Prediction is extremely difficult," Danish physicist Niels Bohr is said to have remarked, "especially about the future."[3] This droll inside joke alludes to prediction at the present moment because of the odd behavior of particles on a subatomic scale.[4] But the dangers of making categorical predictions about the future are self-evident. We cannot see what's ahead for aviation because technological development itself is largely an accidental process. While it is likely that today's jetliners may someday look as antiquated as the DC-3 does to us now, we certainly cannot predict how human thought will flow, what connections will be made, and which ideas will serve future generations.

Our modern jet transports have tremendous pluses in terms of performance and safety. They're also highly efficient. Despite being an order of magnitude faster, newer jetliners are more fuel-efficient than most economy cars in the amount of fuel consumed per passenger seat.

Nevertheless, today's "tube and wing" designs—so called because they have discrete fuselages to house payload and discrete wings to provide lift—do have limitations that frustrate aeronautical engineers. One is that their fuselages generate no lift; they contribute only skin friction that almost doubles the airplane's total aerodynamic drag.

This leads aviation futurists to contemplate *blended wing-body* concepts. In a BWB airplane, a flattened and broadened fuselage that provides its own lift merges seamlessly into the wings, creating an airframe that is considerably more fuel efficient and quieter than today's airplanes.

There are many problems with BWBs that will tend to limit them to lower speeds, lower altitudes, and shorter ranges. They also may not be as inherently stable and may have difficulty meeting safety requirements for takeoff or emergency evacuation.

For airlines, getting passengers to fly aboard BWB airliners would be a very hard sell because the seats would be spread across a wide, roughly triangular floor area with few if any windows. Passengers seated far outboard in this wide cabin would be so far from the airplane's roll axis that normal banked turns would subject them to large vertical translations through space. Depending on the direction of the turn, the airplane's bank would feel either like falling through an open trapdoor or rising suddenly skyward in a Ferris wheel.

Consequently, BWBs are more likely to find military and cargo applications than passenger use. Here their structural-weight efficiencies, payload advantages, greater fuel efficiency, avoidance of environmental emissions at very high altitudes, and exceptional quiet could all be put to productive use.

William Thomson (1824–1907), an Irish-born physicist and mathematician of the nineteenth century, was knighted as Lord Kelvin in Scotland for his contributions to electricity, thermodynamics, and other fields (the Kelvin scale of absolute temperature is named after him). Near the end of a long life, this octogenarian made a prescient prediction just as—unbeknownst to him—the airplane was being invented.

"Young man," Lord Kelvin told a youthful acquaintance, "if you live to be as old as I am, you will be able to breakfast in New York and dine in London—nothing can now prevent the development of the aeroplane in a few years."[5]

Creative humans of both genders and all backgrounds have always dreamed of flying. What will today's young thinkers see and do on this inspiring front in their lifetimes? Whatever it is, it will sing to the human soul just as it has for thousands of years. As Orville put it, "Wilbur and I could hardly wait for morning to come to get at something that interested us. *That's* happiness!"[6]

NOTES

1 CONCEPTION: The Thinker and the Dreamer

1. Charles H. Gibbs-Smith, *Sir George Cayley, 1773–1857* (London: HMSO, 1968), 11.

2. George Cayley, "On Aerial Navigation" (part 1 of 3), *A Journal of Natural Philosophy, Chemistry and the Arts* 24 (1809), 164.

3. Gibbs-Smith, *Sir George Cayley, 1773–1857*, 6.

4. Charles H. Gibbs-Smith, *Sir George Cayley's Aeronautics, 1796–1855* (London: HMSO, 1962), 18.

5. Cayley, "On Aerial Navigation," part 1, 164.

6. Exchange recounted by Cayley's great-granddaughter, an eyewitness to the events; Gibbs-Smith, *Sir George Cayley, 1773–1857*, 21.

7. "Airplane Invention Was Delayed 60 Years," *Fort Worth Press*, October 11, 1961.

8. Application for act of incorporation, read before Parliament by Mr. Roebuck, Frederick Marriott's MP, March 24, 1843.

2 BIRTH: Wilbur, Orville, and the World

1. Orville Wright, *How We Invented the Airplane: An Illustrated History*, ed. Fred C. Kelly (New York: Dover, 1988), 5.

2. Ibid., 84–85.

3. James Tobin, *To Conquer the Air: The Wright Brothers and the Great Race for Flight* (New York: Free Press, 2003), 41.

4. Tom D. Crouch, *The Bishop's Boys: A Life of Wilbur and Orville Wright* (New York: W. W. Norton, 1985), 55.

5. Orville Wright, *How We Invented the Airplane*, 81 (appendix reprint of "The Wright Brothers' Aeroplane" by Wilbur and Orville Wright, *Century Magazine*, September 1908).

6. Ibid.

7. Wilbur Wright to Ohio Society of New York, January 10, 1910; Marvin W. McFarland, ed., *The Papers of Wilbur and Orville Wright* (New York: McGraw-Hill, 1953), 2: 978.

8. Katharine Wright to Henry Haskell, May 18, 1925; Katharine Wright Haskell Papers, Western Historical Manuscript Collection, University of Missouri at Kansas City (microfilm).

9. Tobin, *To Conquer the Air*, 44.

10. Charles H. Gibbs-Smith, *The Invention of the Aeroplane, 1799–1909* (London: Faber and Faber, 1965), 25.

11. Wilbur Wright to Octave Chanute, May 13, 1900; in McFarland, *Papers of Wilbur and Orville Wright*, 1:4.

12. Wilbur Wright to Smithsonian Institution, May 30, 1899, in McFarland, *Papers of Wilbur and Orville Wright*, 1:15.

3 CONFIGURATION: Shapes and Ideas

1. *Washington Post*, October 8, 1903.

2. Edward Jablonski, *Sea Wings: The Romance of the Flying Boats* (New York: Doubleday, 1972), 1.

3. Henry S. Villard, *Contact! The Story of the Early Birds* (New York: Thomas Y. Crowell, 1968), 63.

4. Ibid., 47.

5. Ibid.

6. Ibid, 71.

4 FUSELAGE: Of Drums and Dragonflies

1. "Come Josephine in My Flying Machine" (New York: Shapiro Music, 1910), composed by Fred Fischer with lyrics by Alfred Bryan.

2. This mention appears in Article IV of the Versailles Treaty, which specified the war matériel to be surrendered to the victorious Allies.

5 WINGS, PART I: From Box Kites to Bridges

1. Igor Sikorsky quote, circa 1935, paraphrased by Sergei I. Sikorsky in e-mail to author, September 24, 2007.

2. Lawrence Hargrave, "Paper on Aeronautical Work," *Journal and Proceedings of the Royal Society of New South Wales* 29 (1895), 47.

3. Octave Chanute, *Progress in Flying Machines* (New York: American Engineer and Railroad Journal, 1894).

4. Ibid., 218.

5. Ibid., 231.

6. Francis Wenham, "On Aerial Locomotion and the Laws by Which Heavy Bodies Impelled Through Air Are Sustained," as reprinted in James Means, ed., *The Aeronautical Annual of 1895* (Boston: W. B. Clarke, 1894), 84.

7. Ibid., 82–113, title page.

8. J. Laurence Pritchard, "Francis Herbert Wenham, Honorary Member, 1824–1908: An Appreciation of the First Lecturer to the Aeronautical Society," *Journal of the Royal Aeronautical Society* 62 (August 1958), 580.

9. Theodore W. Fuller, *San Diego Originals* (Pleasant Hills, CA: California Profiles Publications, 1987), 23, 25.

10. Orville Wright, *How We Invented the Airplane: An Illustrated History*, ed. Fred C. Kelly (New York: Dover, 1988), 84.

11. Wilbur Wright, letter to his father, October 2, 1902.

6 WINGS, PART II: Cloud-Cutting Cantilevers

1. Wealthy pioneer U.S. aviator John Moisant had actually designed, built, and tested a one-of-a-kind airplane made entirely of metal in 1909. Constructed of aluminum, that experimental single-seater reportedly hopped rather than flew.

2. The boundary layer is a very thin layer of air immediately over a wing that, because of friction between the air molecules and the wing's surface, has a lower velocity than the overall airflow over the wing. The boundary layer flows smoothly from front to back during normal flight (laminar flow), but at high angles of attack with insufficient airspeed it no longer has sufficient energy to remain laminar all the way to the trailing edge. As a result, boundary-layer separation (turbulent airflow) begins on the wing; this aft point of detachment progresses forward on the wing as the resultant aerodynamic stall deepens.

3. Fly-by-wire jet transports have a programmed yaw-damper function in their computerized flight-control systems rather than a discrete yaw-damper system.

8 FLIGHT CONTROLS: The Chariot's Reins

1. Charles H. Gibbs-Smith, *The Invention of the Aeroplane, 1799–1909* (London: Faber and Faber, 1965), xiv.

2. *Paris Herald*, August 9, 1908.

3. Henry S. Villard, *Contact! The Story of the Early Birds* (New York: Thomas Y. Crowell, 1968), 54.

4. Charles H. Gibbs-Smith, *The Rebirth of Aviation* (London: HMSO, 1974), 286–87.

5. Tom D. Crouch, *The Bishop's Boys: A Life of Wilbur and Orville Wright* (New York: W. W. Norton, 1985), 368.

6. The tips of the *Éole's* wings could be tilted up or down but not forward or aft for effective roll control.

7. Crouch, *The Bishop's Boys*, 167.

8. Louis Mouillard, "The Empire of the Air: An Ornithological Essay on the Flight of Birds," *Annual Report of the Smithsonian Institution, 1893*, 397.

9. Orville Wright, *How We Invented the Airplane: An Illustrated History*, ed. Fred C. Kelly (New York: Dover, 1988), 83.

10. James Tobin, *To Conquer the Air: The Wright Brothers and the Great Race for Flight* (New York: Free Press, 2003), 151.

11. Orville Wright, *How We Invented the Airplane*, 19.

12. Amos I. Root, "Our Homes," *Gleanings in Bee Culture*, January 1, 1905, 36.

13. Ibid.

14. Orville Wright, *How We Invented the Airplane*, 86.

15. In November 1906, Alberto Santos-Dumont added octagonal ailerons to the outer wing bays of the 14-*bis*; while this technically gave his machine three-axis control, in practice it remained only marginally controllable.

16. Charles H. Gibbs-Smith, *The Wright Brothers: A Brief Account of Their Work, 1899–1911* (London: HMSO, 1963), 26.

9 FLIGHT DECK: Cockpits for Aerial Ships

1. Tom D. Crouch, *The Bishop's Boys: A Life of Wilbur and Orville Wright* (New York: W. W. Norton, 1985), 285.

2. Walter S. Ross, *The Last Hero: Charles A. Lindbergh* (New York: Harper & Row, 1968), 105.

3. Ibid.

4. Harry F. Guggenheim, *The Seven Skies* (New York: G. P. Putnam's Sons, 1930), 151.

5. Ibid.

6. Lowell Thomas and Edward Jablonski, *Doolittle: A Biography* (New York: Doubleday, 1976), 95.

7. Ibid., 102.

8. "Blind Plane Flies 15 miles and Lands," *New York Times*, September 25, 1929.

10 AERO PROPULSION: Prometheus Is Pushing

1. George Cayley, "On Aerial Navigation" (part 1 of 3), *A Journal of Natural Philosophy, Chemistry and the Arts* 24 (1809): 164.

2. Howard S. Wolko, *In the Cause of Flight: Technologists of Aeronautics and Astronautics* (Washington, D.C.: Smithsonian Institution Press, 1981), 52.

3. Orville Wright, *How We Invented the Airplane: An Illustrated History*, ed. Fred C. Kelly (New York: Dover, 1988), 85.

4. Ibid.

5. The world's first rotary engine was a small three-cylinder automobile engine that American F. O. Farwell developed in the latter 1890s.

6. Rotary engines were generally not equipped with throttles because the cylinders had to keep spinning fast for adequate cooling. However, some rotaries did have a split air/fuel throttle that was tricky to use. Others had provision for shutting off the ignition to certain cylinders for partial power reduction as an aid to landing. Even when these provisions were available, pilots often ignored them in favor of the all-or-nothing blip switch.

11 LANDING GEAR: Shoes, Canoes, and Carriage Wheels

1. James Tobin, *To Conquer the Air: The Wright Brothers and the Great Race for Flight* (New York: Free Press, 2003), 355.

2. Martin's was one of many early patents awarded by the U.S. government for gear-retraction concepts. The first ever issued to an American was granted to F. McCarroll on November 7, 1915, four years after he applied for it. Another in 1917 went to Princeton student Charles Hampton Grant for what he ineptly termed a "collapsible gear." Grant later participated in the design of the Dayton-Wright XPS-1.

12 PASSENGER CABIN: Voyaging Aloft

1. "Jane Eads Lands on Coast, First Air Passenger: Girl Reporter Tells Story of Crossing Continent in Boeing Plane," *Chicago Herald and Examiner*, July 2, 1927.

2. Ibid.

3. Ibid.

4. Ibid.

13 SYSTEMS INTEGRATION: Making Flying Safer

1. Raymonde de Laroche was the world's first licensed woman pilot.

14 TODAY'S STATE OF THE ART: The Boeing 787 Dreamliner

1. James Burke, *Connections* (New York: Simon & Schuster, 2007), xii.

2. As quoted by Boeing Historical Archives, Bellevue, Washington.

POSTSCRIPT: Tomorrow's Wings

1. While most military aircraft also use kerosene, turbine engines can burn a broad range of fuels, and the military sometimes uses higher distillates.

2. Commercial jets operate at the top of the troposphere and into the stratosphere. The former extends from ground level to about 6 miles (10 kilometers) above the earth's surface, where the latter begins.

3. Niels Bohr as quoted in James Burke, *Connections* (New York: Simon & Schuster, 2007), xii.

4. Bohr is alluding to the Heisenberg uncertainty principle, a quantum mechanics verity.

5. Henry S. Villard, *Contact! The Story of the Early Birds* (New York: Thomas Y. Crowell, 1968), 3.

6. Tom D. Crouch, *The Bishop's Boys: A Life of Wilbur and Orville Wright* (New York: W. W. Norton, 1985), 228.

ACKNOWLEDGMENTS

The author is first and foremost indebted to Elisabeth Dyssegaard, executive editor of Smithsonian Books/HarperCollins and a good friend, for initiating this project. After our previous enjoyable collaboration, Elisabeth asked me what I wanted to write for her next. To my surprise, I heard myself describing this book, a project I didn't know was within me. Without her belief, guidance, and encouragement, this tale might not have been told.

Many individuals have helped me with this unconventional look at flight. Valuable insights and corrections were generously provided by Ph.D. historians Tom Crouch, Bob van der Linden, Peter L. Jakab, and Dick Hallion of the National Air and Space Museum, Smithsonian Institution, in Washington, DC. NASM curator Dorothy Cochrane likewise provided expert assistance at every turn.

For insights about the future of commercial aviation in general and the blended-wing body in particular, I owe thanks to Dr. Thurai Rahulan of the University of Salford, United Kingdom, for his generous and very patient explanations. Dr. John McMasters, Boeing technical fellow and affiliate professor of aeronautics at the University of Washington, likewise shared helpful insights about the past, present, and future of airplane design.

Fortunately for aviation researchers like me, the Boeing Company maintains the largest corporate archives in the aerospace industry. Boeing Historical Services professionals Michael Lombardi and Thomas Lubbesmeyer in Seattle, Washington, and Pat McGinnis at the former Douglas Aircraft archives in Southern California, provided ongoing research help. Mike Lombardi, a fine historian in his own right, also reviewed and suggested improvements to this manuscript.

The author is equally grateful to many other organizations, entities, and individuals. For example, David Craddock, Peter Williams, and Val Gregory, of the Royal Society of New South Wales, helped me properly represent Australian pioneer Lawrence Hargrave. Aviation authority Carroll Gray, creator of the Flying Machines Web site (http://www.flyingmachines.org), also helped, albeit indirectly, by promoting a broad understanding of flight's early pioneers and their contributions.

Boundless gratitude goes to the dedicated professionals who provided this book's inspiring imagery. Melissa Keiser, chief photo archivist at the National Air and Space Museum, Smithsonian Institution in Washington, DC, lent her enthusiastic support to the betterment of the project, as did her colleagues Jessamyn Lloyd and Kate Igoe. So too did Meredith Downs and Amy Heidrick, photo archivist and lead photo archivist respectively at the Museum of Flight in Seattle. Katherine Williams, archivist extraordinaire of the museum's Dahlberg Center for Military Aviation History, likewise turned up many rare photos. Additional imagery came from Tom Lubbesmeyer and Mary Kane at Boeing, Connie Moore at NASA, Derek Pedley at Air Team Images, and other sources. For this bountiful help I remain forever grateful.

Finally, I must express my most heartfelt appreciation to Kate Antony, editorial assistant, designer Jessica Shatan Heslin, and copyeditor Sue Warga at HarperCollins for their contributions to this publication, both in its physical realization and as a downloadable e-book.

PHOTO CREDITS

SOURCE	PAGES (NEG. OR REF. NUMBERS)
Author's collection	[64]
Air Team Images	[198] (50324), [280] (45367)
Boeing	[80] (boe247d), [84] (k10905), [134] (bw65589), [193] (6087b), [194] (c8n), [196] (p8817), [197] (p8620), [226] (7117b), [236] (k63642-07), [261] (c8p), [264] (1773b), [266] (eads2), [269] (hs640), [275] (p8300), [276] (p8657), [296] (k64021-02), [302] (k64262-03), [303] (k63450-07)
Getty Images	frontispiece (2642581)
Library of Congress	19, [107], [111], [164], [206]
Museum of Flight, Seattle	Cover, [65], [66], [68], [69], [73], [74], [81], [87], [115], [118], [120], [127], [129], [166], [209], [229], [245], [256], [257], [260], [270]
NASA	[308]
National Air and Space Museum, Smithsonian Institution	5 (2004-20825), 8 (2004-20823), 11 (89-12480), 13 (96-16071), 17 (93-7192), 25 (74-10606), 29 (2003-12097), 30 (87-17029), 31a (2004-20822), 31b (73-9000), 33 (2003-12100), 43 (85-18303), 45a (78-15850), 45b (87-6034), 46 (97-16613), 47 (75-7731), 49 (2008-3768), [50] 78-14976, [52] (2008-3762), [54] (2008-3773), [62] (2008-3767), [63] (89-1882), [67] (86-13488), [71] (2001-11622), [72] (2008-3774), [75] (76-2425), [76] (73-8002), [78] (75-12127), [83] (77-5256), [88] (89-226), [92] (2008-3769),

	[93] (2006-6055), [96] (2008-3770), [99] (76-17158), [108] (2003-12979), [113] (78-15849), [*121*] (77-2701), [133] (97-17485), [139] (85-3940), [145] (93-12783), [146] (88-8142), [147] (95-2473), [150] (2005-17747), [168] (85-18299), [169] (2002-16644), [175] (2008-3775), [183] (2008-5599), [188] (81-878), [190] (75-14896), [191] (83-410), [214] (2003-12303), [222] (83-12593), [231] (87-2330), [242] (2008-3763), [243] (2008-3764), [254] (2008-3761), [255] (83-16525), [263] (75-7026), [271] (2005-3678)
National Portrait Gallery, London	2

INDEX

Page numbers in *italics* indicate illustrations.